Jean René Boudreau
2013

A Guide
to
Plane Algebraic Curves

© 2011 by
The Mathematical Association of America (Incorporated)

Library of Congress Catalog Card Number 2011932374

Print Edition ISBN 978-0-88385-353-5
Electronic Edition ISBN 978-1-61444-203-5

Printed in the United States of America

Current Printing (last digit):
10 9 8 7 6 5 4 3 2

The Dolciani Mathematical Expositions
NUMBER FORTY-SIX

MAA Guides # 7

A Guide

to

Plane Algebraic Curves

Keith Kendig
Cleveland State University

Published and Distributed by
The Mathematical Association of America

DOLCIANI MATHEMATICAL EXPOSITIONS

Committee on Books
Frank Farris, *Chair*

Dolciani Mathematical Expositions Editorial Board
Underwood Dudley, *Editor*
Jeremy S. Case
Rosalie A. Dance
Tevian Dray
Thomas M. Halverson
Patricia B. Humphrey
Michael J. McAsey
Michael J. Mossinghoff
Jonathan Rogness
Thomas Q. Sibley

The DOLCIANI MATHEMATICAL EXPOSITIONS series of the Mathematical Association of America was established through a generous gift to the Association from Mary P. Dolciani, Professor of Mathematics at Hunter College of the City University of New York. In making the gift, Professor Dolciani, herself an exceptionally talented and successful expositor of mathematics, had the purpose of furthering the ideal of excellence in mathematical exposition.

The Association, for its part, was delighted to accept the gracious gesture initiating the revolving fund for this series from one who has served the Association with distinction, both as a member of the Committee on Publications and as a member of the Board of Governors. It was with genuine pleasure that the Board chose to name the series in her honor.

The books in the series are selected for their lucid expository style and stimulating mathematical content. Typically, they contain an ample supply of exercises, many with accompanying solutions. They are intended to be sufficiently elementary for the undergraduate and even the mathematically inclined high-school student to understand and enjoy, but also to be interesting and sometimes challenging to the more advanced mathematician.

1. *Mathematical Gems,* Ross Honsberger
2. *Mathematical Gems II,* Ross Honsberger
3. *Mathematical Morsels,* Ross Honsberger
4. *Mathematical Plums,* Ross Honsberger (ed.)
5. *Great Moments in Mathematics (Before 1650),* Howard Eves
6. *Maxima and Minima without Calculus,* Ivan Niven
7. *Great Moments in Mathematics (After 1650),* Howard Eves
8. *Map Coloring, Polyhedra, and the Four-Color Problem,* David Barnette
9. *Mathematical Gems III,* Ross Honsberger
10. *More Mathematical Morsels,* Ross Honsberger
11. *Old and New Unsolved Problems in Plane Geometry and Number Theory,* Victor Klee and Stan Wagon
12. *Problems for Mathematicians, Young and Old,* Paul R. Halmos
13. *Excursions in Calculus: An Interplay of the Continuous and the Discrete,* Robert M. Young
14. *The Wohascum County Problem Book,* George T. Gilbert, Mark Krusemeyer, and Loren C. Larson
15. *Lion Hunting and Other Mathematical Pursuits: A Collection of Mathematics, Verse, and Stories by Ralph P. Boas, Jr.,* edited by Gerald L. Alexanderson and Dale H. Mugler
16. *Linear Algebra Problem Book,* Paul R. Halmos
17. *From Erdős to Kiev: Problems of Olympiad Caliber,* Ross Honsberger
18. *Which Way Did the Bicycle Go? ... and Other Intriguing Mathematical Mysteries,* Joseph D. E. Konhauser, Dan Velleman, and Stan Wagon
19. *In Pólya's Footsteps: Miscellaneous Problems and Essays,* Ross Honsberger

20. *Diophantus and Diophantine Equations*, I. G. Bashmakova (Updated by Joseph Silverman and translated by Abe Shenitzer)
21. *Logic as Algebra*, Paul Halmos and Steven Givant
22. *Euler: The Master of Us All*, William Dunham
23. *The Beginnings and Evolution of Algebra*, I. G. Bashmakova and G. S. Smirnova (Translated by Abe Shenitzer)
24. *Mathematical Chestnuts from Around the World*, Ross Honsberger
25. *Counting on Frameworks: Mathematics to Aid the Design of Rigid Structures*, Jack E. Graver
26. *Mathematical Diamonds*, Ross Honsberger
27. *Proofs that Really Count: The Art of Combinatorial Proof*, Arthur T. Benjamin and Jennifer J. Quinn
28. *Mathematical Delights*, Ross Honsberger
29. *Conics*, Keith Kendig
30. *Hesiod's Anvil: falling and spinning through heaven and earth*, Andrew J. Simoson
31. *A Garden of Integrals*, Frank E. Burk
32. *A Guide to Complex Variables* (MAA Guides #1), Steven G. Krantz
33. *Sink or Float? Thought Problems in Math and Physics*, Keith Kendig
34. *Biscuits of Number Theory*, Arthur T. Benjamin and Ezra Brown
35. *Uncommon Mathematical Excursions: Polynomia and Related Realms*, Dan Kalman
36. *When Less is More: Visualizing Basic Inequalities*, Claudi Alsina and Roger B. Nelsen
37. *A Guide to Advanced Real Analysis* (MAA Guides #2), Gerald B. Folland
38. *A Guide to Real Variables* (MAA Guides #3), Steven G. Krantz
39. *Voltaire's Riddle: Micromégas and the measure of all things*, Andrew J. Simoson
40. *A Guide to Topology*, (MAA Guides #4), Steven G. Krantz
41. *A Guide to Elementary Number Theory*, (MAA Guides #5), Underwood Dudley
42. *Charming Proofs: A Journey into Elegant Mathematics*, Claudi Alsina and Roger B. Nelsen
43. *Mathematics and Sports*, edited by Joseph A. Gallian
44. *A Guide to Advanced Linear Algebra*, (MAA Guides #6), Steven H. Weintraub
45. *Icons of Mathematics: An Exploration of Twenty Key Images*, Claudi Alsina and Roger B. Nelsen
46. *A Guide to Plane Algebraic Curves*, (MAA Guides #7), Keith Kendig

MAA Service Center
P.O. Box 91112
Washington, DC 20090-1112
1-800-331-1MAA   FAX: 1-301-206-9789

Preface

This book was written as a friendly introduction to plane algebraic curves.

It's for...

- Mathematicians who never took a course on algebraic curves, or took one years ago and have forgotten most of it.

- Students who are curious about algebraic curves and would like an easy-to-read account of what it is and what its major highlights are.

- Anyone taking an elementary course on algebraic curves. This book can serve as a useful companion, supplying perspective and concrete examples to flesh out abstract concepts.

- Outsiders who have heard that algebraic geometry is useful in attacking an increasingly wide range of applied problems and want an entry point that doesn't require an extensive mathematical background.

What this book is, and what it isn't.

- **What it is.** This book emphasizes geometry and intuition, and the presentation is kept concrete. Learning about plane algebraic curves provides a foundation for going on to higher dimensional algebraic geometry. We work mainly over the complex numbers, where results are beautifully unified and consistent. You'll find an abundance of pictures and examples to help develop your intuition, so basic to understanding and asking fruitful questions. The book covers the highlights of the elementary theory which for some could be an end in itself, and for others an invitation to investigate further, including algebraic geometry and more general methods.

- **What it isn't.** This is not a "Theorem, Proof, Corollary" book. Proofs, when given, are mostly sketched, some in more detail, but typically with less. We often include references to texts that provide further discussion.

WHAT ARE THE PREREQUISITES FOR THIS BOOK?

- The rudiments of coffee cup and donut topology.

- Some basic complex analysis, including Cauchy-Riemann equations, complex-analytic functions, meromorphic functions, and Laurent expansions.

- The definitions of field, field isomorphism, algebraic extension of a field, integral domain, ideal and prime ideal.

WHY SHOULD I BE INTERESTED, IN ALGEBRAIC CURVES?

Since about 1990, algebraic curves and algebraic geometry have undergone explosive growth. Computer algebra software has made getting around in algebraic geometry much easier. Algebraic curves and geometry are now being applied to areas such as cryptography, complexity and coding theory, robotics, biological networks, and coupled dynamical systems. Algebraic curves were used in Andrew Wiles' proof of Fermat's Last Theorem, and to understand string theory, you need to know some algebraic geometry. There are other areas on the horizon for which the concepts and tools of algebraic curves and geometry hold tantalizing promise. This introduction to algebraic curves will be appropriate for a wide segment of scientists and engineers wanting an entrance to this burgeoning subject.

A BIT OF PERSPECTIVE.

This book follows the traditional approach of working over the complex numbers, an approach that played a large role in setting up the subject and remains a natural way to enter it. In the early part of the 20th century, this found grand expressions in works on algebraic functions by Appell and Goursat, as well as by Hensel and Landsberg. Dover reprints of [Bliss] and [Coolidge] give a good perspective of a slightly later period. We've taken the somewhat more contemporary approach found in [Walker] or [Fulton], but for concreteness, we do almost everything over the complex numbers.

The Book's Story Line . . .

- In Chapter 1 we visit a gallery of algebraic curves in the real plane. The examples show the surprisingly wide range of possible behavior, and a section on Designer Curves further drives home the point by providing principles for creating an even broader array of user-defined curves. The apparent jungle of possibilities leads to a basic question: *Where are the nice theorems?*

- A fundamental truth emerges in Chapter 2: to get nice theorems, algebraic curves must be given enough living space. For example, important things can happen at infinity, and points at infinity are beyond the reach of the real plane. We use a squeezing formula to shrink the entire plane down to a disk, allowing us to view everything in it. This picture leads to adjoining points at infinity, and in one stroke all sorts of exceptions then melt away. We enhance the reader's intuition through pictures showing what some everyday curves look like after squeezing them into a disk.

- Chapter 3 continues the quest originating from Chapter 1: *Where are the nice theorems?* Once again, the answer lies in giving algebraic curves additional living space—in this case we expand from the real numbers to the complex. Working over them, together with points added at infinity, we arrive at one of the major highlights of the book, Bézout's theorem. This is one of the most underappreciated theorems in mathematics, and it represents an outstandingly beautiful generalization of the Fundamental Theorem of Algebra. Our proof uses the resultant—a double-edged sword which itself is one of the most underappreciated tools in mathematics. We use one edge in Chapter 1, and the other in Chapter 3.

- Chapter 4 continues our quest. In Chapter 1 we met curves that are connected, and others that are not. There are curves of pure dimension, and others that aren't. From what seems like a nearly hopeless situation, Chapters 2 and 3 lay a foundation for establishing one of the most important and satisfying topological properties of algebraic curves: a curve defined by an irreducible polynomial in its complex numbers-based habitat is always topologically connected, and is a real 2-manifold with finitely many points identified to finitely many points. We even know the surface must be orientable. In a sense that we'll make precise, "most" algebraic curves are both irreducible and require no point identifications, so topologically nearly every algebraic curve is an orientable 2-manifold like a sphere, donut, the surface of a bagel with two holes in it, and so on. We derive a remarkably simple formula for its genus in terms of the defining polynomial's degree.

- Chapter 5 trains a magnifying glass on some of the results seen so far. The prettiest and simplest of them statistically hold for 100% of algebraic curves, but nonetheless there exist curves—many with very simple defining polynomials—that bend, twist and contort so much that in order to fit in the plane, they must have self intersections and/or kinks. Such points are rare (accounting for their name "singularities"), but rare or not, questions arise:

  - *What do curves look like around singularities?*
  - *Are some singularities easily understood, while others are more complicated?*
  - *How is their number and type related to the amount of twisting and contorting of the curve?*
  - *For a curve with singularities, what happens to Bézout's theorem?*
  - *For a curve with singularities, what happens to that remarkably simple genus formula?*
  - *Can you transform a curve with singularities into a curve without singularities?*

  Chapter 5 provides answers. In fact, the answer to the last question is "yes," and the actual theorem once again highlights algebraic curves' need for enough living space: in transforming a curve with singularities to one without, we may need to grant the curve an extra dimension, allowing it to live in a complex three-dimensional world instead of in just two.

- In Chapter 6, a large cluster of seemingly disparate facts about curves, discovered over several generations of mathematicians, are gathered into a commutative diagram. Earlier generations — the ancient Greeks — carried out the first exhaustive study of any subject in mathematics: algebraic curves of degree 2. They were known then and are known today as conics. The simplest curves after conics are those curves of degree 3 that have no singular points, and this means each is topologically a torus. By focusing on such a specific genre of curve, many more detailed results ensue. Their study turns out to be deep and rewarding,

Preface

and the story is still incomplete. An appropriate commutative diagram pulls together many of their basic properties and links up the three concepts of irreducible curve, its function field, and its Riemann surface. Each of the three determines the other two up to an appropriate notion of equivalence. Finally, in concluding the book, we shake hands with an important idea: transporting elementary complex-variable theory from a nonsingular curve of genus 0 (this corresponds to the typical first "one complex variable" course) to a compact Riemann surface of any genus. This represents a surprising change in flavor of the study. A good number of pictures are provided to enhance intuition.

Many thanks to . . .

- Don Albers, who suggested writing this book.

- Underwood Dudley, whose keen writing sense tightened up the exposition throughout.

- Dolciani editorial board members Jeremy Case, Rosalie Dance, Tevian Dray, Thomas Halverson, Patricia Humphrey, Michael McAsey, Michael Mossinghoff, Jonathan Rogness, and Thomas Sibley, who critiqued the final draft.

- Basil Gordon, whose many suggestions, both mathematical and expositional, greatly improved the book.

- Richard Scott, who provided helpful feedback on early outlines of the book.

- Ivan Soprunov, who read the entire manuscript and checked the examples for correctness.

- Beverly Ruedi, whose technical expertise has been an inspiration to me. It was Bev who led me to computer drawing software, and I was able to create all the illustrations in this book using either Adobe Illustrator or importing plots from Maple and then applying Illustrator.

- Carol Baxter, who skillfully led this opus through to publication.

Cleveland, Ohio   Keith Kendig

Contents

Preface vii

1 A Gallery of Algebraic Curves 1
    1.1   Curves of Degree One and Two 1
    1.2   Curves of Degree Three and Higher 4
    1.3   Six Basic Cubics 7
    1.4   Some Curves in Polar Coordinates 7
    1.5   Parametric Curves 12
    1.6   The Resultant 13
    1.7   Back to an Example 15
    1.8   Lissajous Figures 16
    1.9   Morphing Between Curves 18
    1.10 Designer Curves 22

2 Points at Infinity 29
    2.1   Adjoining Points at Infinity 30
    2.2   Examples 33
    2.3   A Basic Picture 35
    2.4   Basic Definitions 37
    2.5   Further Examples 40

3 From Real to Complex 45
    3.1   Definitions 46
    3.2   The Idea of Multiplicity; Examples 47
    3.3   A Reality Check 52
    3.4   A Factorization Theorem for Polynomials in $\mathbb{C}[x, y]$ . . . 54
    3.5   Local Parametrizations of a Plane Algebraic Curve 55
    3.6   Definition of Intersection
           Multiplicity for Two Branches 56
    3.7   An Example 57

    3.8   Multiplicity at an Intersection Point of Two Plane Algebraic Curves 58
    3.9   Intersection Multiplicity Without Parametrizations 59
    3.10 Bézout's theorem 63
    3.11 Bézout's theorem Generalizes the Fundamental Theorem of Algebra 69
    3.12 An Application of Bézout's theorem: Pascal's theorem . . 71

4 Topology of Algebraic Curves in $\mathbb{P}^2(\mathbb{C})$ 75
    4.1   Introduction 75
    4.2   Connectedness 76
    4.3   Algebraic Curves are Connected 77
    4.4   Orientable Two-Manifolds 84
    4.5   Nonsingular Curves are Two-Manifolds 86
    4.6   Algebraic Curves are Orientable 87
    4.7   The Genus Formula 88

5 Singularities 93
    5.1   Introduction 93
    5.2   Definitions and Examples 94
    5.3   Singularities at Infinity 97
    5.4   Nonsingular Projective Curves 97
    5.5   Singularities and Polynomial Degree 99
    5.6   Singularities and Genus 102
    5.7   A More General Genus Formula 110
    5.8   Non-Ordinary Singularities 111
    5.9   Further Examples 115
    5.10 Singularities versus Doing Math on Curves 117
    5.11 The Function Field of an Irreducible Curve 118
    5.12 Birational Equivalence 119
    5.13 Examples of Birational Equivalence 121
    5.14 Space-Curve Models 127
    5.15 Resolving a Higher-Order Ordinary Singularity 130
    5.16 Examples of Resolving an Ordinary Singularity 131
    5.17 Resolving Several Ordinary Singularities 137
    5.18 Quadratic Transformations 138

6 The Big Three: C, K, S 143
    6.1   Function Fields 145
    6.2   Compact Riemann Surfaces 146
    6.3   Projective Plane Curves 152

6.4	f_1, f_2, f: Curves and Function Fields	153
6.5	g_1, g_2, g: Compact Riemann Surfaces and Curves	155
6.6	h_1, h_2, h: Function Fields and Compact Riemann Surfaces	156
6.7	Genus	156
6.8	Genus 0	157
6.9	Genus One	158
6.10	An Analogy	166
6.11	Equipotentials and Streamlines	170
6.12	Differentials Generate Vector Fields	174
6.13	A Major Difference	175
6.14	Divisors	179
6.15	The Riemann-Roch theorem	182

Bibliography 185

Index 189

About the Author 193

CHAPTER 1

A Gallery of Algebraic Curves

A great way to learn new mathematics is to work with examples. That's how we start. This chapter consists mostly of examples of algebraic curves in the real plane. A plane algebraic curve is defined to be the locus, or set of zeros, of a polynomial in two Cartesian variables with real coefficients. This may sound pretty special, but a surprisingly large number of familiar curves are exactly of this type. For example, many polar coordinate curves — lemniscates, limaçons, all sorts of roses, folia, conchoids—are algebraic, as are many curves defined parametrically, such as Lissajous figures and the large assortment of curves obtained by rolling a circle of rational radius around a unit circle. Nearly all the curves the ancient Greeks knew are algebraic. So are many curves mechanically traced out by linkages.

We begin this chapter with very simple algebraic curves, those defined by first and second degree polynomials. We then turn to curves of higher degree.

1.1 Curves of Degree One and Two

Definition 1.1. The *degree of a monomial* $x^m y^n$ is $m + n$. The *degree of a polynomial* $p(x, y)$ is the largest degree of its terms. The *degree of a plane algebraic curve* C is the degree of the lowest-degree polynomial defining C.

Notation. In this book, we denote the set of all solutions of $p(x, y) = 0$ by $C(p(x, y))$ or by just $C(p)$.

1

Degree One

The general form of a polynomial of degree one is $Ax + By + C$, where not both A and B are zero. Its zero set is a line, and conversely any line in $\mathbb{R}^2$ is the zero set of a polynomial of degree one.

Geometrically, two distinct points in the plane determine a unique line. This has an algebraic translation: in $Ax + By + C = 0$, not both A and B are zero, so assume that $A \neq 0$. Dividing by it gives $x + \beta y + \gamma = 0$. Substituting into it the coordinates of two points in the plane produces two linear equations in the β and γ. If the points are distinct, the equations are linearly independent and therefore uniquely determine values for β and γ, thus defining a line in the plane.

Degree Two

The general form of a polynomial of degree two is

$$Ax^2 + Bxy + Cy^2 + Dx + Ey + F$$

where not all of A, B and C are zero. Its zero set is a conic that can be non-degenerate, degenerate, or the empty set, and any conic in $\mathbb{R}^2$ is the zero set of some degree-two polynomial. The non-degenerate conics are ellipses (including circles), parabolas and hyperbolas, while degenerate ones include the empty set (defined by $x^2 + 1 = 0$, for example), two crossing lines (example: $xy = 0$) or parallel lines (as in $x^2 - 1 = 0$) or two copies of the same line (example: $x^2 = 0$). We call two coincident copies of the same line a *double line*.

As with a line, a certain number of points uniquely determine a conic. To see what this number is, replay the algebraic argument above: divide

$$Ax^2 + Bxy + Cy^2 + Dx + Ey + F = 0$$

by one of A, B and C to get, for example,

$$x^2 + \beta xy + \gamma y^2 + \delta x + \epsilon y + \phi. \tag{1.1}$$

If five distinct points in the plane are chosen so that no more than three are collinear, then substituting them into (1.1) gives a linearly independent system of five equations that uniquely determines β through ϕ, and therefore a conic. If three points are collinear, the conic is degenerate since it must contain a line. (An appreciation for how five points determine a conic can be

1.1. Curves of Degree One and Two

gained by experimenting with the five-points conic routine in the geometry software package Cabri.)

If in the general degree-two polynomial the linear terms are absent and $F = -1$, then the conic is defined by

$$Ax^2 + Bxy + Cy^2 = 1,$$

and is symmetric about the origin. The discriminant $B^2 - 4AC$ then indeed discriminates, telling us that in the real plane the conic is either empty or

an ellipse if $B^2 - 4AC < 0$;
two parallel lines if $B^2 - 4AC = 0$;
a hyperbola if $B^2 - 4AC > 0$.

What effect does adding the linear part $Dx + Ey$ have on the conic $Ax^2 + Bxy + Cy^2 = 1$? If the discriminant is nonzero, then this will shift the conic, and uniformly magnify it (zoom in) if it's an ellipse, or zoom in or out if it's a hyperbola. It does not change the shape of either conic. If the discriminant is zero, then $Ax^2 + Bxy + Cy^2 = 1$ defines two parallel lines, and adding $Dx + Ey$ can change them into a parabola. An example is $A = E = 1$ and $B = C = D = 0$. For details, see Chapter 9 of [Kendig 1].

Most calculus and pre-calculus books choose equations to make the conics "nice," and this usually leaves misleading impressions. As we swim around in the sea of all conic sections, what do we actually encounter? We can mimic such a tour by taking a series of snapshots as we move about, a photo corresponding to randomly selecting real values for $A, \ldots, F$. Dividing an equation through by a nonzero number doesn't change the zero set, so without loss of generality, we can assume their values are in the interval $(-1, 1)$. Random choices mean none of $A, \ldots, F$ are ever exactly zero. Here are some things we will and won't see:

- We never see a parabola.

- We never encounter a non-empty degenerate conic.

- We never see a conic having principal axes parallel to the x- and y-axes, as in the standard forms of an ellipse or hyperbola.

- Often we see what appears to be a parabola, but by zooming out far enough, we will see that either the curve closes up to form an ellipse or we encounter another branch, showing that the curve is a hyperbola.

- In the case of $Ax^2 + Bxy + Cy^2 = 1$, we can look at $B^2 - 4AC = 0$ as defining a surface within the cube $(-1, 1) \times (-1, 1) \times (-1, 1)$. We never land on this surface, which is the boundary between points (A, B, C) in the cube corresponding to hyperbolas on one side and ellipses or the empty set on the other. The cube is divided into three pieces:

  i. The part where $B^2 - 4AC > 0$, corresponding to hyperbolas;

  ii. The part where $B^2 - 4AC < 0$ and both A and C are positive, corresponding to ellipses;

  iii. The part where $B^2 - 4AC < 0$ and both A and C are negative, corresponding to the empty set.

By finding the volumes of these pieces, we can find the probability that randomly picking a point from the cube produces an ellipse. The cube is divided into eight unit cubes, one in each octant, and only two of these eight contribute volume corresponding to ellipses. Writing the boundary surface as $z^2 = 4xy$ and using symmetry leads to a probability of

$$\frac{2}{8} \int_{z=0}^{1} \int_{x=\frac{z^2}{4}}^{1} \left(1 - \frac{z^2}{4x}\right) dx\, dz.$$

This turns out to be (only!)

$$\frac{31 - 6\ln 2}{144} \approx 18.6397\%.$$

The probability of getting the empty set is the same, $\approx 18.6397\%$, and the probability of a hyperbola is approximately

$$100\% - 37.279\% = 62.721\%.$$

For further reading, [Kendig 1] is an accessible account of many ideas in this book for second-degree curves — that is, conics.

1.2 Curves of Degree Three and Higher

Degree Three

The subtlety and complexity of curves having degree n increase rapidly with n. Curves of degree one fall into just one class: lines. Curves of degree two can be put into four main classes: ellipses, parabolas, hyperbolas, and degenerate cases. (In the complex setting, the degenerate cases are two lines, either different or coincident.) But by degree three, categorizing becomes so

1.2. CURVES OF DEGREE THREE AND HIGHER

nontrivial that to this day there is no one classification considered "best" or most natural. There exist useful classifications based on various criteria, one being Newton's analytic classification. He massages the general two-variable cubic

$$Ax^3 + Bx^2y + Cxy^2 + Dy^3 + Ex^2 + Fxy + Gy^2 + Hx + Jy + K$$

into one of four special forms in which either y, y^2, xy or $xy^2 + \alpha y$ is set equal to the pure one-variable cubic $ax^3 + bx^2 + cx + d$. There are 78 cases in all; Newton found 72 of them. (See [B-K], section 2.5 for a nice discussion.)

The statistical game we played for curves of degree two can be run for curves of degree three. By randomly choosing real values for $A, \ldots, K$ in the general two-variable cubic, we encounter certain shapes of real cubics again and again, while others appear less frequently or very rarely. The six

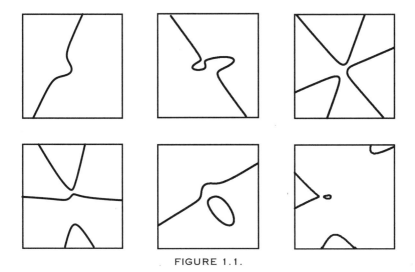

FIGURE 1.1.

snapshots of cubic curves in Figure 1.1 suggest a few possible shapes. They are arranged from most to least frequently encountered, going from top left to bottom right. By far the most common is the shape at the top left, showing a single bump. Sometimes it's more S-shaped, as in the next picture. Together, these account for about 70% of randomly chosen curves. The next two are variants of each other, each consisting of three separate branches, and will be seen about 20% of the time. A bump with an island occurs perhaps 5% of the time, and the last, three branches plus an island arises rarely, less than 1% of the time. These percentages are very approximate. There are other shapes that occur even more rarely.

Since we have ten coefficients $A, \ldots, K$, dividing through by any one of $A, \ldots, D$ leaves nine degrees of freedom. However, it is possible to select nine distinct points that do not uniquely determine a cubic:

Example 1.1. Figure 1.2 shows nine points, with two different cubics passing through them: the graph of $y = 3x(x-1)(x+1)$ and the graph of $x = 3y(y-1)(y+1)$.

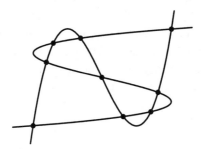

FIGURE 1.2.

Higher Degrees

We have seen that the number of degrees of freedom for a curve of degree n is

$2 = 3 - 1$ for degree 1
$5 = 6 - 1$ for degree 2
$9 = 10 - 1$ for degree 3.

The numbers 3, 6 and 10 are called *triangular* because in the following arrangement of polynomial forms of increasing degree, the number of terms of degree $\leq n$ is like triangle areas, starting at the top: 1, 3, 6, 10, ... :

$$A$$
$$Ax + By$$
$$Ax^2 + Bxy + Cy^2$$
$$Ax^3 + Bx^2y + Cxy^2 + Dy^3$$
$$\cdots$$

The number of degrees of freedom enjoyed by a curve of degree n is one less than the number of terms in its general equation. Remembering that the sum of the first m natural numbers $1 + 2 + 3 + \cdots + m$ is $\frac{(m)(m+1)}{2}$, we see

1.3 Six Basic Cubics

that the number of degrees of freedom in a curve of degree n is

$$\frac{(n+1)(n+2)}{2} - 1 = \frac{n^2 + 3n}{2}.$$

The first few of these numbers are $2, 5, 9, 14, 20, 27, 35, 44, 54, 65, \ldots$.

1.3 Six Basic Cubics

There are certain cubics that play a special role in studying algebraic curves, since they illustrate a variety of basic concepts. The graph of $y = p(x)$, where $p(x)$ is a typical cubic in x, can be pushed up and down to five essentially different positions that reflect the nature of its roots — all real, not all real, repeated or distinct. In the left column of Figure 1.3 we get all five positions by starting the graph low and pushing it upward. Any graph of $y = p(x)$ in the left column corresponds to the graph of $y^2 = p(x)$ in the right. Replacing y by y^2 makes all the curves in the right column symmetric with respect to the x-axis.

The same process can be applied to $p(x) = x^3$, in which the polynomial in x has three equal roots. This yields our sixth basic cubic, $y^2 = x^3$, and is an example of a *cusp curve*, depicted in Figure 1.4.

1.4 Some Curves in Polar Coordinates

Rectangular versus Polar Coordinates

Draw circles of latitude and semi-circles of longitude on a sphere. In a small neighborhood around a point on the equator, the latitudes and longitudes closely approximate the horizontal and vertical lines of a rectangular coordinate system. But at the opposite extremes, at the north or south pole, the latitudes and longitudes look like the circles and rays of a polar coordinate system. In this sense, rectangular and polar coordinates are opposites of each other. This behavior extends to many familiar curves in rectangular versus polar coordinates, as we'll discover in a moment.

Algebraic versus Not Algebraic

Many plane curves may be algebraic, yet are not presented as the zero set of a polynomial $p(x, y)$ in rectangular coordinates. For example, the curve might be given by an equation in polar coordinates, or by a pair of parametric equations, or traced out by some mechanical linkage or as the path

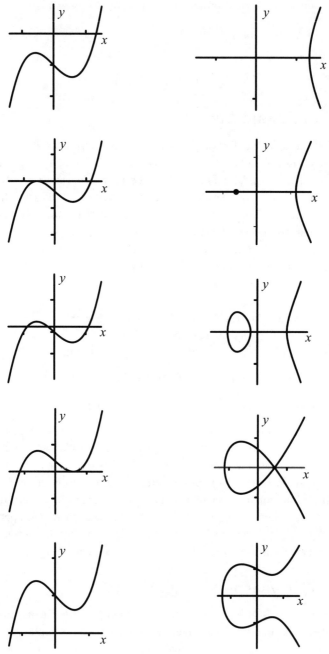

FIGURE 1.3.

1.4. Some Curves in Polar Coordinates

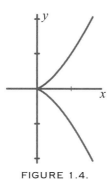

FIGURE 1.4.

of a point on one curve as it rolls along another one. If such a "roulette" is algebraic, then by definition it is the zero set of some polynomial $p(x, y)$. It just may not be obvious what that polynomial is.

There's also an abundance of non-algebraic curves, such as graphs of trigonometric functions, so the question arises: is there an easy way to tell algebraic curves from non-algebraic ones? Here's a partial test: if there is a line in the plane intersecting the curve in infinitely many discrete points, then the curve is not algebraic. For example, the graph of $y = \cos x$ is not algebraic since the x-axis intersects it in infinitely many discrete points. Virtually all graphs of trigonometric functions fail to be algebraic. In $\mathbb{R}^2$, this test is only sufficient. For example, most lines cross the graphs of $y = e^x$ or $\ln x$ in infinitely many distinct points, but only in $\mathbb{C}^2$, not $\mathbb{R}^2$. These graphs are not algebraic curves.

The Oppositeness Idea

We now look at a kind of oppositeness between the behavior of a curve defined a polynomial equation $p(x, y) = 0$ versus its polar counterpart obtained by replacing y by r and x by θ and plotting in $\mathbb{R}^2$ as in elementary analytic geometry. A few examples give the flavor. One of the simplest of all algebraic curves is defined by $y = x$. The polar counterpart of this equation is $r = \theta$, defining an Archimedean spiral. This spirals outward infinitely many times. Any line intersects it in infinitely many discrete points, so the spiral isn't algebraic. More generally, if $p(x)$ is a polynomial of positive degree, then the graph of $y = p(x)$ is algebraic while the corresponding graph $r = p(\theta)$ isn't, because the size of $r = p(\theta)$ increases without bound as θ approaches $\pm\infty$. This means, for example, that the curve $r = p(\theta)$ intersects the x-axis in the infinitely many distinct points having polar coordinates $(r, \theta) = (p(n\pi), n\pi)$. On the other hand, the graph of

$y = \cos x$ is not algebraic, while its polar brother $r = \cos \theta$ defines a circle — algebraic indeed.

This phenomenon does not always hold, but it often points us in the right direction. As just a few examples, the polar equations for such familiar curves as the conics, lemniscates, cardioids, cissoids, limaçons, cochleoids, all sorts of roses, as well as the Witch of Agnesi, all involve trigonometric functions in their definitions, yet all are algebraic. That means their polar equations can all be converted to polynomial equations in rectangular coordinates. Sometimes the conversion is straightforward, as with $r = \cos \theta$, where we can write $r^2 = r \cos \theta$ and then substitute $x^2 + y^2$ for r^2 and x for $r \cos \theta$ to obtain $x^2 + y^2 = x$. Other times the conversion entails more work. Here's an example.

Example 1.2. The polar equation $r = \cos 4\theta$ defines a rose having eight petals. Intersecting the curve with a line can give us some basic information. Let's say $p(x, y)$ has degree n. Substituting the parametrization $\{x = at + b, y = ct + d\}$ of a general line into $p(x, y)$, yields a polynomial in t of degree n, $p(at + b, ct + d)$. The line intersects $C(p(x, y))$ at those points corresponding to values of t for which $p(at + b, ct + d)$ is zero. Therefore the line should intersect the curve in n points. We may not see all n in the real plane since some may be complex or some zeros may be repeated. But if we can find a line intersecting the curve (in this case, the rose) in m points, then we know the polynomial $p(x, y)$ defining it must have degree at least m. Figure 1.5 depicts the eight-petal rose. We see that the horizontal line intersects five petals in two points each, for a total of 10 points of intersection. Therefore the degree of p must be at least 10.

FIGURE 1.5.

For the conversion, use a double-angle formula to go from $\cos 4\theta$ to a trigonometric function of 2θ, and again to go to a trigonometric function

1.4. Some Curves in Polar Coordinates

of θ. In the end, $r = \cos 4\theta$ becomes

$$r = \cos^4\theta - 6\cos^2\theta\sin^2\theta + \sin^4\theta.$$

Replace $\cos\theta$ by $\frac{x}{\sqrt{x^2+y^2}}$ and $\sin\theta$ by $\frac{y}{\sqrt{x^2+y^2}}$. We need to be careful with r, since for θ between 0 and 2π, half the time the polar value r is negative. Replacing r by $+\sqrt{x^2+y^2}$ gives the four loops pointing in the directions of the x- and y-axes, and they have the equation

$$\sqrt{x^2+y^2} = \frac{x^4}{(x^2+y^2)^2} - 6\frac{x^2 y^2}{(x^2+y^2)^2} + \frac{y^4}{(x^2+y^2)^2}.$$

That is,

$$(x^2+y^2)^{\frac{5}{2}} = x^4 - 6x^2 y^2 + y^4.$$

Replacing r by $-\sqrt{x^2+y^2}$ yields the other four loops defined by

$$-(x^2+y^2)^{\frac{5}{2}} = x^4 - 6x^2 y^2 + y^4.$$

Squaring both sides introduces no additional solutions and therefore gives the following as the equation of the complete rose:

$$(x^2+y^2)^5 = (x^4 - 6x^2 y^2 + y^4)^2.$$

We were lucky — the horizontal line in our rose picture predicted a polynomial of degree 10 or greater, and the polynomial turns out to have degree exactly 10.

The polynomial equation defining the $4 \cdot 2 = 8$-petal rose can be written in a suggestive way. Denote $x + iy$ by z. It is straightforward to check that the equation $(x^2+y^2)^5 = (x^4 - 6x^2 y^2 + y^4)^2$ can be written as

$$(x^2+y^2)^{4+1} = [\Re(z^4)]^2,$$

where $\Re$ denotes the real part of a complex number. With a bit more work, it can be verified that for n even, the $2(n+1)$-degree equation

$$(x^2+y^2)^{n+1} = [\Re(z^n)]^2 \qquad (1.2)$$

defines the $2n$-petal rose $r = \cos n\theta$ in $\mathbb{R}^2$, plotted as in analytic geometry. In distinction to n even, notice that for n odd, $r = \cos n\theta$ yields only n petals. Also, for odd n, something else different happens: the equation in (1.2) can be written as

$$(x^2+y^2)^{n+1} - \Re(z^n)^2$$
$$= \left((x^2+y^2)^{\frac{n+1}{2}} + \Re(z^n)\right)\left((x^2+y^2)^{\frac{n+1}{2}} - \Re(z^n)\right) = 0,$$

so now the defining polynomial is reducible and defines the union of two different roses, each with n petals.

One more observation: there's no n such that $r = \cos n\theta$ or $r = \sin n\theta$ draws a rose whose petal count is double an odd number, such as $2, 6, 10, \ldots$. We discuss roses further on pp. 131–136.

1.5 PARAMETRIC CURVES

Many curves presented parametrically turn out to be algebraic, one obvious example being $\{x = at + b, y = ct + d\}$ which defines a line. But what about $\{x = q_1(t), y = q_2(t)\}$, where q_1 and q_2 are polynomials? Specifically, just picking an example from the air, what about the curve defined by

$$\{x = 3t^2 + t + 1, \ y = t^4 - 4t^3 - 5\}?$$

If it's algebraic, we should be able to eliminate t and get a polynomial $p(x, y)$ whose zero set is the parametrized curve. It is not obvious how to get rid of t, but it can be done, and the polynomial $p(x, y)$ turns out to be

$$x^4 - 56x^3 - 18x^2 y + 72x^2 - 84xy + 81y^2 - 580x + 899y + 2523.$$

This was derived not through incredible cleverness or endless toil, but rather by using a powerful tool, the *resultant*. It took less than a minute to type the appropriate command into Maple, which then carried out the algebra in much less than a second. We present this extremely useful tool in a moment.

Even if q_1 and q_2 are not polynomials but, say, trigonometric functions, the curve can be algebraic. For example,

$$\{x = \cos t, \ y = \sin t\}$$

defines a circle since squaring each of $\cos t$ and $\sin t$ and adding eliminates t and yields $x^2 + y^2 = 1$. But consider the slightly fancier curve

$$\{x = \cos 3t, \ y = \sin 2t\}. \tag{1.3}$$

Here again it's not obvious how to eliminate t, but as we'll see on p. 17, the resultant again comes to our aid, producing a polynomial whose zero set is the curve. Therefore this fancier curve is algebraic.

1.6 THE RESULTANT

We use two single-variable polynomials of degree m and n to give the basic idea of what the resultant is and what it does. All these ideas easily generalize. Let

$$q_1(t) = a_0 t^m + a_1 t^{m-1} + \cdots + a_m = 0,$$
$$q_2(t) = b_0 t^n + b_1 t^{n-1} + \cdots + b_n = 0.$$

A major question the resultant answers is "Do $q_1(t)$ and $q_2(t)$ share a common zero?" One might suggest finding the zeros of each polynomial and checking to see if any are the same. However, if one polynomial has degree five or higher it may be possible only to approximate the zeros, so we could not be sure they agree exactly. The resultant is the determinant of a square matrix whose entries are the polynomials' coefficients a_i and b_j, and it will give a simple, direct and exact answer to our question.

To see the idea behind the resultant, suppose q_1 and q_2 share a zero c, which may be complex. The polynomials then have the common factor $(t-c)$, so we can write

$$q_1(t) = (t-c)r(t) \quad \text{and} \quad q_2(t) = (t-c)s(t),$$

where
$$r(t) = \alpha_0 t^{m-1} + \alpha_1 t^{m-2} + \cdots + \alpha_{m-1}$$

and
$$s(t) = \beta_0 t^{n-1} + \beta_1 t^{n-2} + \cdots + \beta_{n-1},$$

with α_0, β_0 nonzero. The assumption that q_1 and q_2 have a common factor $(t-c)$ implies that

$$s q_1 - r q_2 = 0$$

since both sq_1 and rq_2 equal $(t-c)rs$. Conversely, a sufficient condition for q_1 and q_2 to have a common factor is that there exist s with $\deg(s) < n$ and r with $\deg(r) < m$ so that $sq_1 - rq_2 = 0$. This is because any factor of q_1 must appear among the factors of rq_2. They cannot all occur in r because the degree of r is less than the degree of q_1, so at least one of them must occur among the factors of q_2.

We now harness the power of linear algebra. We write our necessary and sufficient condition $sq_1 - rq_2 = 0$ as

$$(\beta_0 t^{n-1} + \beta_1 t^{n-2} + \cdots + \beta_{n-1}) \cdot (a_0 t^m + a_1 t^{m-1} + \cdots + a_m) -$$
$$(\alpha_0 t^{m-1} + \alpha_1 t^{m-2} + \cdots + \alpha_{m-1}) \cdot (b_0 t^n + b_1 t^{n-1} + \cdots + b_n) = 0. \quad (1.4)$$

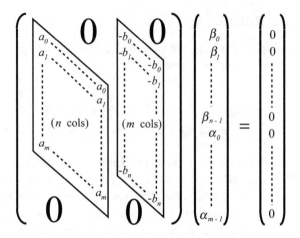

FIGURE 1.6.

Figure 1.6 shows this written in matrix form.

If the criterion $sq_1 - rq_2 = 0$ is to have a nontrivial solution, then the determinant of the square matrix must be zero. By multiplying some of the matrix columns by -1 and transposing, we can write its determinant as in Figure 1.7.

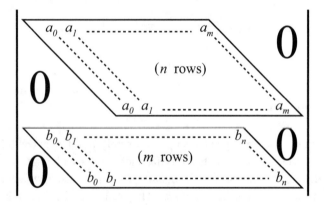

FIGURE 1.7.

The determinant is called the *resultant* of the polynomials

$$a_0 t^m + a_1 t^{m-1} + \cdots + a_m \text{ and } b_0 t^n + b_1 t^{n-1} + \cdots + b_n,$$

and the resultant is zero if and only if the two polynomials have a common zero.

The coefficients a_i and b_j are constants, but they can also be polynomials. For example, $p(x,t)$ and $q(x,t)$ can be regarded as polynomials in t with coefficients in $\mathbb{R}[x]$. Then the resultant isn't a constant, but rather a polynomial in x. This extends to any number of variables. For example, $p(x,y,t)$ and $q(x,y,t)$ could be regarded as polynomials in t with coefficients in $\mathbb{R}[x,y]$, and the resultant would be a polynomial in x and y.

We now introduce some basic notation.

Notation. To indicate that we are regarding p and q as polynomials in t, we write
$$\mathrm{resultant}(p,q,t)$$
and call it *the resultant of p and q with respect to t*, with corresponding meanings for $\mathrm{resultant}(p,q,x)$ and $\mathrm{resultant}(p,q,y)$.

Comment 1.1. In the above, $\mathbb{R}$ can be replaced by $\mathbb{C}$ — that is, the numerical coefficients can be complex, not just real.

1.7 Back to an Example

Let's revisit the curve on p. 12 parametrized by
$$\{\, x = 3t^2 + t + 1,\; y = t^4 - 4t^3 - 5 \,\}.$$

We claimed that this curve is algebraic, but actually eliminating t to get a polynomial in x and y seemed like an all-but-impossible task. We can now let the resultant do its magic. Rewrite the equations in the parametrization as
$$3t^2 + t + (-x+1)t^0 = 0,$$
$$t^4 - 4t^3 - (y+5)t^0 = 0.$$

Any specific point (x_0, y_0) is on the curve if and only if $3t^2 + t + (-x_0 + 1)$ and $t^4 - 4t^3 - (y_0 + 5)$ (which are now polynomials in t with constant coefficients) have a common factor — that is, if and only if there's a value of t making them both zero and therefore satisfying the curve's two parametric equations at (x_0, y_0). We know this happens exactly when
$$\mathrm{resultant}(3t^2 + t - x_0 + 1, t^4 - 4t^3 - y_0 - 5, t) = 0.$$

Since (x_0, y_0) is arbitrary, we conclude that any point (x, y) is on the curve when $\mathrm{resultant}(3t^2 + t - x + 1, t^4 - 4t^3 - y - 5, t)$ is zero. This is a polynomial in x and y whose zero set is the curve! That polynomial is exactly what we want, and is the determinant of a 6×6 matrix with entries in $\mathbb{R}[x, y]$:

$$\begin{vmatrix} 3 & 1 & (-x+1) & 0 & 0 & 0 \\ 0 & 3 & 1 & (-x+1) & 0 & 0 \\ 0 & 0 & 3 & 1 & (-x+1) & 0 \\ 0 & 0 & 0 & 3 & 1 & (-x+1) \\ 1 & -4 & 0 & 0 & (-y-5) & 0 \\ 0 & 1 & -4 & 0 & 0 & (-y-5) \end{vmatrix}.$$

Maple or Mathematica saves us computing by hand and gives

$$x^4 - 56x^3 - 18x^2y + 72x^2 - 84xy + 81y^2 - 580x + 899y + 2523,$$

just as claimed on p. 12. The resultant method shows more generally that any curve parametrized by polynomials $q_1(t), q_2(t)$ is algebraic.

1.8 Lissajous Figures

A Lissajous figure is a curve traced out by two sinusoidal motions

$$x = A \cos mt, \quad y = B \sin(nt - \phi).$$

The shape is often created in physics demos using a laser beam and two tuning forks having a mirror attached to each. The beam bouncing off the first mirror is aimed at the second, which then reflects the beam to a viewing screen. The forks are oriented so the laser beam picks up both sinusoidal motions. The curve closes and continues to trace over itself.

The curve in (1.3) on p. 12 parametrized by

$$\{x = \cos 3t, \quad y = \sin 2t\}$$

is one such example, depicted in Figure 1.8. It is symmetric about the origin and lives in the square with the four vertices $(\pm 1, \pm 1)$. Is this parametric curve algebraic? That is, is it possible to eliminate t, arriving at a polynomial $p(x, y)$ whose zero set is the curve? The answer is yes. We start by using double- and triple-angle formulas to express the trigonometric functions in terms of $\cos t$:

$$x = \cos 3t = 4\cos^3 t - 3\cos t,$$

$$y = \sin 2t = 2 \cos t \sin t = \pm 2 \cos t \sqrt{1 - \cos^2 t}.$$

1.8. Lissajous Figures

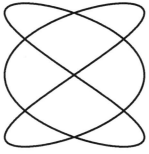

FIGURE 1.8.

The curve is symmetric about the origin, so both sides of the last equation can be squared without adding any new points to the curve. This yields

$$y^2 = 4\cos^2 t - 4\cos^4 t.$$

Regard $\cos t$ as a parameter; call it T. Parametric equations for the Lissajous figure may now be written

$$\{4T^3 - 3T - x = 0, \ -4T^4 + 4T^2 - y^2 = 0\}.$$

To eliminate T, take

$$\text{resultant}(4T^3 - 3T - x, -4T^4 + 4T^2 - y^2, T).$$

Using Mathematica or Maple and dividing the result by a constant gives

$$p(x, y) = 16y^6 + 4x^4 - 24y^4 - 4x^2 + 9y^2.$$

In the real plane, the zero set of this polynomial is precisely the Lissajous figure in Figure 1.8. As a partial check, notice that selecting values for x yields a polynomial in y of degree six. For $x_0 \in [-1, 1]$, the vertical line $x = x_0$ intersects the Lissajous figure in six points, counting any double points as two points. Likewise, selecting any $y_0 \in [-1, 1]$ yields a polynomial in x of degree four, and the horizontal line $y = y_0$ correspondingly intersects the Lissajous figure in four points.

What about general Lissajous figures parametrized by

$$\{x = A\cos mt, \ y = B\sin(nt - \phi)\} \ ?$$

Are they algebraic for integers m and n? We can again use multiple-angle formulas (which are all just the result of applying addition formulas enough times) to transform the original parametric equations into polynomial equations in terms of x, y and $T = \cos t$. Then take the resultant with respect to T, arriving at a single polynomial $p(x, y)$. The only caveat is that Lissajous figures created in a physics lab are bounded, while if the Lissajous figure is "degenerate" in the sense that the plot doubles back and changes direction in retracing itself, then the zero set in $\mathbb{R}^2$ of $p(x, y)$ extends the Lissajous figure to an unbounded figure. An example is the Lissajous figure $\{x = \cos t, y = \cos 2t\}$. Eliminating t gives $p(x, y) = y + 1 - 2x^2$, so its zero set is the parabola $y + 1 = 2x^2$. But as t runs through real values in the parametrization, only the part of this parabola within the square $[-1, 1] \times [-1, 1]$ gets plotted. At either end of the real curve, the plotted point "bounces back" — reverses direction — as t steadily increases. (The remainder of the real parabola is traced out when t takes on all pure imaginary values.) Adding a small phase shift to make the parametric equations read, say, $\{x = \cos(t + 0.1), y = \cos 2t\}$, removes the degeneracy and produces the situation depicted by the solidly drawn part in Figure 1.9. The

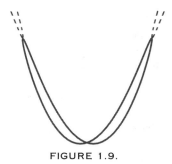

FIGURE 1.9.

dashed part is meant to represent two branches going off to infinity in $\mathbb{C}^2$. These are not visible in the real plane, but make the figure is unbounded in $\mathbb{C}^2$. Figure 5.2 on p. 101 depicts two fancier degenerate examples.

1.9 Morphing Between Curves

Two algebraic curves defined by polynomials $p(x, y)$ and $q(x, y)$ can morph into each other, with all intermediate curves being algebraic. A variable $\alpha \in [0, 1]$ can serve as the morphing parameter. As α increases from 0 to 1, the linear combination $(1 - \alpha)p + \alpha q$ morphs from p to q, and the zero sets of these linear combinations morph from the zero set of p to that of q. Geometrically, some morphings are obvious. An example is the horizontal line $p(x, y) = y$ changing into the vertical line $q(x, y) = x$: the

1.9. Morphing Between Curves

horizontal line simply rotates to the vertical one, the intermediate lines being $(1-\alpha)y + \alpha x = 0$. But morphing can bridge any curve to any other, and sometimes the sequence of pictures is quite astonishing.

Example 1.3. Figure 1.10 shows the 8-petal rose in Figure 1.5 on p.10 morphing into the Lissajous figure in Figure 1.8. All intermediate curves are algebraic.

FIGURE 1.10.

So far, we've chosen α to be in the interval $[0, 1]$, but α can just as well be any real number. In a sense, the morphing movie corresponding to α increasing from 0 to 1 can be extended indefinitely into the past and future. The set of all such morphed or blended curves $(1-\alpha)p(x, y) + \alpha q(x, y)$ ($\alpha \in \mathbb{R}$) has two basic properties:

Property 1. The intersection $C(p) \cap C(q)$ belongs to every blended curve, in that for each $\alpha \in \mathbb{R}$, the corresponding curve must pass through each point of $C(p) \cap C(q)$. Reason: Suppose P is in both $C(p)$ and $C(q)$. Then $p(P) = q(P) = 0$, so $(1-\alpha)p(P) + \alpha q(P) = 0$, meaning P belongs to $C((1-\alpha)p(x, y) + \alpha q(x, y))$.

Property 2. The totality of all blends of $C(p)$ and $C(q)$ covers the plane $\mathbb{R}^2$. To eliminate the degenerate case where one blend is the zero polynomial defining not a curve but the whole plane, assume that p and q are not scalar multiples of each other. To show that any $Q \in \mathbb{R}^2$ is in some blend curve, note that this is trivially true if $Q \in C(p) \cap C(q)$. Therefore let Q be any point not in $C(p) \cap C(q)$, say $q(Q) \neq 0$, and blend p and q via $p + \lambda q$, $\lambda \in \mathbb{R}$. Then for some λ, $p + \lambda q = 0$. Since p and q are not scalar multiples of each other, $C(p + \lambda q)$ is not $\mathbb{R}^2$.

1. A Gallery of Algebraic Curves

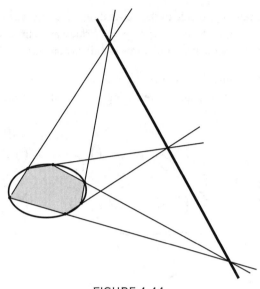

FIGURE 1.11.

The following example uses both of these properties.

Example 1.4. Figure 1.11 shows a hexagon inscribed in an ellipse. Any two opposite sides of the hexagon extend to a pair of lines intersecting in a point. There are three such intersection points, and remarkably, they always turn out to be collinear. This is an instance of "Pascal's theorem" saying that collinearity holds for any hexagon inscribed in any ellipse. A picture guide to the theorem's proof can be given as a series of blended curves. This appears in Figure 1.12. An ellipse is shown with hexagon vertices on it numbered clockwise, and we see three alternate extended hexagon sides forming the union C_1 of three lines shown in the top left picture. The bottom right picture depicts the analogous curve C_2 consisting of the other three extended hexagon sides. The eight frames depict the morphing from C_1 to C_2 as α increases from 0 to 1. C_1 and C_2 intersect in nine points; notice that every intermediate cubic does in fact contain all nine points, as the above Property 1 promises. Property 2 tells us that for some α, the blend curve contains a point on the ellipse different from any of the six vertex points. It's exactly at that α when the magic happens: the blend curve then becomes reducible, splitting up into an ellipse plus the line containing those three intersection points. We fill in details of the proof in Chapter 3, p. 71.

Any two algebraic curves can be bridged via morphing, and in fact the initial and final curves need not be algebraic — that is, p and q need not

1.9. Morphing Between Curves

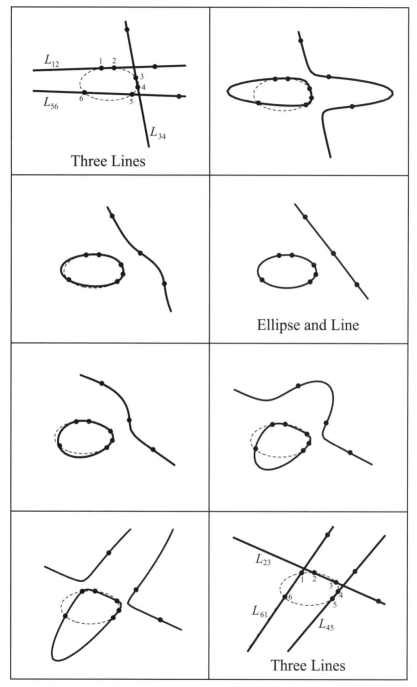

FIGURE 1.12.

be polynomials. For example, the graph of a sine function can morph to an Archimedean spiral, or into a circle. A good way to learn and appreciate morphing is to run computer animations. Typing the command **?animations** in a Maple work sheet leads to examples that the user can easily modify; the morpher α plays the role of time. (After reading Chapter 3, it will be clear that α could equally well run through $\mathbb{C}$ instead of $\mathbb{R}$.)

1.10 Designer Curves

Software packages such as Mathematica and Maple can plot virtually any curve whose equation is typed in, making these packages wonderful at translating algebra to geometry. But what about going the other way, finding equations for curves we visualize? One result pointing in this direction is the graph of the Lagrange interpolating polynomial of degree n passing through $n+1$ points $(x_0, y_0), \cdots, (x_n, y_n)$, the x_i being distinct. For example, there is a unique graph of a cubic $y = a_0 x^3 + a_1 x^2 + a_2 x + a_3$ passing through four such points P_0, P_1, P_2, P_3. A 20th century analogue of this is a segment of a Bézier cubic in which P_0, P_1, P_2, P_3 can be any four points in the plane. The segment passes through P_0 and P_3, starting at P_0 and heading toward P_1. The longer the vector $P_0 P_1$, the more closely the curve tangentially hugs the line through P_0 and P_1. Similarly, at the segment's ending point P_3, the curve segment heads toward P_2, and the longer the vector $P_3 P_2$, the more closely the segment, tangent to the line through P_3 and P_2, hugs that line. If a rubber band is stretched around P_0, P_1, P_2, P_3 to form a convex polygon, the segment is contained within the polygon. More complex shapes are created by taking such Bézier segments to be building blocks and joining them together smoothly. This intuitive method is used in drawing programs such as CorelDraw! and Adobe Illustrator.

In the world of algebraic curves, there are some curve-making principles that can be used to design a wide variety of curves. We obtain a single polynomial whose zero set is, or approximates, the shape we're looking for. In what follows, each bulleted item represents a different principle.

- $C(p) \cup C(q) = C(pq)$.

That is, to get the union of two curves, multiply their defining polynomials. The reasoning is as follows. We have $C(p) \cup C(q) \subseteq C(pq)$ because a point P in $C(p) \cup C(q)$ must be in at least one of $C(p)$ or $C(q)$, so at least one of $p(P)$, $q(P)$ is zero, meaning $p(P)q(P) = 0$. Also, $C(p) \cup C(q) \supseteq C(pq)$ because if $p(P)q(P) = 0$ for a point P,

1.10. Designer Curves

then either $p(P) = 0$ or $q(P) = 0$, since otherwise the product would be nonzero.

Example 1.5. The union of the x- and y-axes is the zero set of xy. Their union with the lines $y = \pm x$ is defined by $xy(y^2 - x^2)$.

Example 1.6. The union of two circles of radii 1 and 2 centered at the origin has equation

$$(x^2 + y^2 - 1)(x^2 + y^2 - 4) = 0.$$

The surface $z = (x^2 + y^2 - 1)(x^2 + y^2 - 4)$ is sombrero-shaped having two rings at height $z = 0$.

Definition 1.2. The *order* of a polynomial p is the lowest degree among its terms. The *initial part* or *lowest-degree part* of p is the polynomial consisting of all terms of degree equal to the order of p.

- In a sufficiently small neighborhood U of the origin, the zero set $C(p) \cap U$ is approximated by the zero set of the lowest-degree part of p.

The zero set of the lowest-degree part consists of finitely many lines, each tangent to the curve at the origin, and is known as the *tangent cone* to the curve at the origin. It's a cone in the sense that it consists of lines through the origin, as in three-space. Here's the idea behind this principle: $r < 1$ implies that $r^{n+1} < r^n$, so for r small enough, the initial part q satisfies $p - q > q$ throughout a disk centered at the origin and having radius r. In fact, $\frac{q}{p-q}$ can be made as small as we wish by taking $r > 0$ sufficiently small.

This principle works just as well around any point P of $C(p)$ — just translate $C(p)$ so that P moves to the origin. Alternatively, expand the polynomial about P.

Example 1.7. In $p = x^2 - y^2 + x^3$, the lowest-degree part is $x^2 - y^2$, so its zero set is given by $y^2 = x^2$ and consists of the two lines $y = \pm x$. The zero set of p is the alpha curve depicted in Figure 1.13. The two dashed lines $y = \pm x$ are tangent to the curve, and the lowest-degree part of p defines the tangent cone to the curve at the origin.

Example 1.8. In the 8-petal rose in Figure 1.5 on p. 10, the lowest-degree part of its defining polynomial is $(x^4 - 6x^2y^2 + y^4)^2$. It can be checked

1. A Gallery of Algebraic Curves

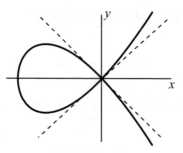

FIGURE 1.13.

that $(x^4 - 6x^2y^2 + y^4)$ is the product of the four factors

$$\left[y \pm \left(1 + \sqrt{2}\right)x\right], \left[y \pm \left(1 - \sqrt{2}\right)x\right].$$

They define four lines tangent to the eight petals. There are two smooth arcs on each side of any tangent line. Each arc has a tangent line at the origin, so for each pair of two tangent arcs there are two tangent lines. This is reflected in the exponent 2 in the lowest-degree part $(x^4 - 6x^2y^2 + y^4)^2$.

Example 1.9. Suppose that we'd like the tangent cone to consist of the x-axis ($y = 0$) together with the two lines $y = \pm x$. Form $y(x^2 - y^2)$, and add a higher-degree term like x^4 so the curve is not the union of three lines, yet has $y(x^2 - y^2)$ as lowest-degree part. At the origin, the tangent cone of the zero set of $y(x^2 - y^2) + x^4$ is indeed the union of these three lines, as in the left sketch in Figure 1.14. By also adding y^4 to get the polynomial $y(x^2 - y^2) + x^4 + y^4$, the entire real figure becomes bounded, as shown on the right.

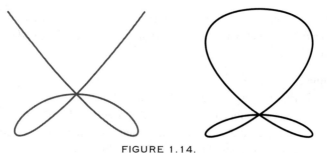

FIGURE 1.14.

Definition 1.3. If the degree of p is m, then the *highest-degree part of p* is what remains of p after all terms of degree less than m are removed.

1.10. Designer Curves

Just as a polynomial's lowest-degree part in an expansion about a point P gives "behavior in the small," a polynomial's highest degree part gives "behavior in the large":

- The zero set of the highest-degree part of a polynomial p looks like the asymptotes to a greatly zoomed-out view of $C(p)$.

Zooming out can be thought of as looking at the viewing plane $\mathbb{R}^2$ from ever greater distances, which amounts to uniformly shrinking the unit of measurement. On a computer screen, plotting a curve within the square having the four vertices $(\pm 10,000, \pm 10,000)$ gives a zoomed-out view, compared to the part of its plot within the square $(\pm 1, \pm 1)$. Any bounded curve will look like a point after zooming out far enough. If the curve is unbounded, then at a point (x_0, y_0) on the curve far from the origin, $p(x_0, y_0)$'s value of zero is attained mainly because tremendously large values of p's terms cancel out. The major portion comes from the highest-degree terms, with terms of lower degree contributing, relatively speaking, very little. The location of (x_0, y_0) is close to a point where only the highest-degree part vanishes.

Example 1.10. The familiar parabola $y - x^2$ is a surprising example because the extremely near-the-origin and extremely far-away views are perpendicular to each other. That is, if we zoom into the origin we see approximately $y = 0$, the x-axis. But if we zoom very far out so the viewing square has side one billion, then for most of the parabola points we see, the ratio of the y-value to the x-value is large, approaching a billion as we approach the edge of the square. Visually, the parabola looks like a vertical ray extending upward from the origin.

Example 1.11. Suppose $q(x)$ is a polynomial of degree greater than 1. If the degree is even, then viewed from a sufficiently great distance the graph of $y = q(x)$ looks like a vertical ray extending either upward or downward from the origin. If the degree is odd, it looks like the entire y-axis.

Example 1.12. Take something wild, like

$$\prod_{k=1}^{30}(y - kx)$$

and add to it any $p(x, y)$ of degree less than 30. The algebraic curve defined by the sum will approach 30 different asymptotes.

- For $p(x, y)$ of degree n, $C(p)$ can be bounded in the real plane by adding $x^{2m} + y^{2m}$ to $p(x, y)$, where $2m > n$.

Let $x^{2m} + y^{2m}$ be the largest-degree part of $p(x, y) + x^{2m} + y^{2m}$. By zooming out far enough, its zero set looks like a point, because the zero set of $x^{2m} + y^{2m}$ is the origin. Actually, the only thing necessary is that the highest-degree part have the origin as its zero set. Adding a term of the form $(x^2 + y^2)^m$ works nicely and sometimes brings about greater symmetry.

Example 1.13. The polynomial defining the 8-petal rose in Figure 1.5 on p. 10 is
$$(x^2 + y^2)^5 - (x^4 - 6x^2y^2 + y^4)^2.$$
The highest-degree part is $(x^2 + y^2)^5$ and it forces the four double lines given by $(x^4 - 6x^2y^2 + y^4)^2$ to curl around and remain bounded.

Example 1.14. The cusp curve $y^3 = -x^3$ pictured on the left in Figure 1.15 is unbounded, but if we add $x^4 + y^4$ to get $y^3 = -x^2 + x^4 + y^4$, the branches turn around and close up. The part of the cusp near the origin is barely affected, and we end up with the teardrop curve on the right in Figure 1.15.

FIGURE 1.15.

We add two mini-principles:

- Replacing x and/or y in $p(x, y)$ by high powers can sometimes make the curve more "angular."

Example 1.15. Replacing the circle's equation $x^2 + y^2 = 1$ by $x^8 + y^8 = 1$ or $x^{10} + y^{10} = 1$ produces a curve looking more like a square with rounded corners. If we push the general "Fermat curve" $x^n + y^n = 1$ to an extreme by choosing $n = 200$, then we get what looks to the unaided eye like a genuine square. (It isn't!)

- If the constant term of $p(x, y)$ is zero, then the curve goes through the origin. Adding a small nonzero constant to the polynomial will move

1.10. Designer Curves

the curve away from the origin. Changing the sign of the constant can force curve-reconnections to take place in opposite ways.

Example 1.16. Adding and subtracting an appropriate $\epsilon > 0$ to the polynomial

$$(x^2 + y^2)^5 - (x^4 - 6x^2 y^2 + y^4)^2$$

defining the 8-petal rose produces the left and right pictures in Figure 1.16.

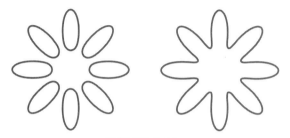

FIGURE 1.16.

Example 1.17. There is an algebraic curve looking like Figure 1.17. How can we find a polynomial creating it? Adding a constant sometimes removes

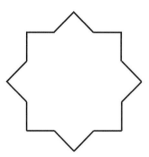

FIGURE 1.17.

part of the curve around the origin. The Fermat curve defined by the polynomial $p(x, y) = x^{200} + y^{200} - 1$ looks like a square, and this "square," rotated by 45°, can be defined by $q = p\left(\left(\frac{x-y}{\sqrt{2}}\right), \left(\frac{x+y}{\sqrt{2}}\right)\right)$. The product pq defines their union, and adding -2 to this product removes some inner material. Here's the curve's polynomial equation:

$$\left(x^{200} + y^{200} - 1\right)\left(\left(\frac{x-y}{\sqrt{2}}\right)^{200} + \left(\frac{x+y}{\sqrt{2}}\right)^{200} - 1\right) = 2.$$

Linkages

Algebraic curves are being increasingly applied to robotics, in which parts of a robot (essentially linked line segments) trace out curves. A central result is Kempe's Universality Theorem stating that every bounded part of an algebraic curve can be generated by an endpoint of some linkage. (See [Abbott] for details and a proof.) A linkage is a finite set of rigid line segments forming a chain or chains in which each endpoint is either fixed or free to rotate around the endpoint of some other segment of the linkage.

Example 1.18. Figure 1.18 depicts a mechanical system that traces out part of an alpha curve. The center of each circle is fixed and any solidly drawn radius is free to turn about the center.

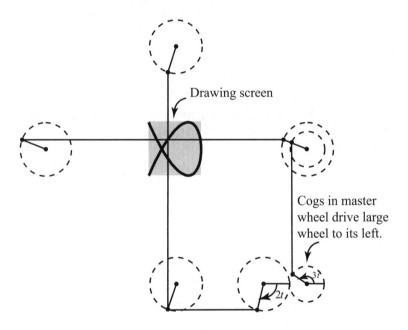

FIGURE 1.18.

CHAPTER 2

Points at Infinity

The examples in the last chapter reveal a wide range of behavior of algebraic curves in the real plane.
- Some are bounded, others are not.
- Some form one piece — that is, they're connected, having just one topological component — while others are not.
- For a curve having two or more topological components, there can be a mixture of bounded and unbounded components. For example, in Figure 1.10 on p. 19, all components are bounded. The middle right graph of Figure 1.3 on p. 8 has a bounded and an unbounded component together. In a hyperbola, both branches are unbounded.
- Even the dimension may not be 1. In the real plane, for example, the locus of $x^2 + y^2 = 0$ consists of the origin which has dimension 0, while $x^2 + y^2 + 1 = 0$ defines the empty set which has dimension -1.
- The curve's dimension can be mixed. For example, in Figure 1.3 on p. 8 we see a cubic having both a one-dimensional component and a zero-dimensional component.
- If the curve's defining polynomial has degree n, there are times when a line intersects the curve in n points, but there are other times when there are fewer than n intersections. As one example, there are always lines completely missing any bounded curve. As another, two distinct lines usually intersect in 1 point, but parallel lines intersect in no points. And the zero set of the degree-two polynomial $x^2 + y^2$ is just a point, so no line intersects the locus in 2 points.

With so much unpredictable and seemingly erratic behavior, you may be wondering:

Where are the nice theorems?

Historically, splintering into many different cases often tells us that we're not looking at a big enough picture. Just think of the exceptions arising in using the quadratic formula to solve quadratic equations if we knew only about nonnegative real numbers! Matters improve if we expand our world to all reals $\mathbb{R}$, but it's when we expand our horizons to the complex numbers $\mathbb{C}$ that solving quadratics becomes beautifully exception-free.

We are about to embark on a similar journey for algebraic curves. As it now stands, our world is too small in two different ways. First, important things happen "at infinity," so we will adjoin points there. In this chapter, we will do this to the real plane, arriving at *the real projective plane* $\mathbb{P}^2(\mathbb{R})$. In this extended plane, *any* two different lines will intersect in exactly one point, no exceptions. Second, the entire landscape of algebraic curves improves tremendously when we work over $\mathbb{C}$ instead of just $\mathbb{R}$. We will take that step in the next chapter, and our definition of $\mathbb{P}^2(\mathbb{R})$ will easily generalize to the complex analogue $\mathbb{P}^2(\mathbb{C})$. At that stage, we will be in just the right position to answer the question *Where are the nice theorems?*

2.1 Adjoining Points at Infinity

To begin, we introduce an example that will motivate much of what we do in this and the next chapter. In $\mathbb{R}^2$, randomly select m lines. Each line is the zero set of a first-degree polynomial. The first "designer principle" on p. 22 tells us that their union is defined by the product of the m first-degree polynomials. The union is therefore an algebraic curve C_1 of degree m. Similarly, select in $\mathbb{R}^2$ n lines to create an algebraic curve C_2 of degree n. Because the lines were selected randomly, **the curves of degree m and n intersect in mn points.** Separately translating or rotating any number of the lines by a sufficiently small amount leaves the number of intersection points unchanged. Perhaps there is a multiplication theorem here.

But let's interfere further with these two curves, rotating a line in one curve so that it becomes parallel to some line in the other curve. During the rotation, the point where the two lines intersect races off to infinity and disappears when the lines become parallel. This "lost" point means the curves now intersect in only $mn - 1$ points. Pushing this idea to an extreme, by suitably rotating lines in C_1 and C_2 so that they're all mutually parallel, we decrease the number of intersection points of C_1 and C_2 to zero! Losing points this way ruins any chances for the suggested multiplication theorem.

Instead of "we lose points," it's more profitable to say "we lose sight of points," since the real problem is that we can't see what's going on that

2.1. ADJOINING POINTS AT INFINITY

far away. There is a simple remedy: shrink $\mathbb{R}^2$ down to a disk, where it will be easy to see what's happening, and then contemplate the problem of disappearing points.

Actually, we've all seen a mapping that does that to $\mathbb{R}$. The principal-value function $y = \arctan x$ maps $\mathbb{R}$ to the interval $(-\pi/2, \pi/2)$. An even simpler function is $x \to \frac{x}{\sqrt{x^2+1}}$, which shrinks $\mathbb{R}$ to $(-1, 1)$. This graph, sketched in Figure 2.1, suggests that the map shrinks distances severely and

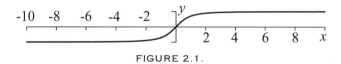

FIGURE 2.1.

unevenly. For example, the image of the unit interval $[0, 1]$ has length about 0.707, while the image of the unit interval $[5, 6]$ just five units away, has length less than 0.006 — less than 1% of .707.

We can apply this shrinking map idea to lines through the origin to map $\mathbb{R}^2$ onto the open unit disk. This map does it:

$$(x, y) \longrightarrow \left(\frac{x}{\sqrt{x^2+y^2+1}}, \frac{y}{\sqrt{x^2+y^2+1}} \right). \tag{2.1}$$

The image of a line L such as $y = 1$ illustrates the shrinking, since the unit vertical distances from L to the x-axis in $\mathbb{R}^2$ decrease under the map. Figure 2.2 shows the image in the open disk of ten lines in $\mathbb{R}^2$ — the y-axis and the nine parallel lines $y = \frac{j}{4}, (j = -4, \ldots, 4)$. For any fixed

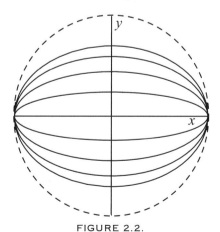

FIGURE 2.2.

$a \in \mathbb{R}$, the images of $y = \pm a$ lie on an ellipse. Its semi-major axis has length 1, and since $(0, a) \in \mathbb{R}^2$ maps to $\left(0, \frac{a}{\sqrt{a^2+1}}\right)$, the semi-minor axis of the ellipse has length $\frac{a}{\sqrt{a^2+1}}$. It is easily checked that the image of any point $(c, \pm a)$ satisfies the equation of this ellipse.

More generally, the image of all mutually parallel lines is a family of ellipses with the same major axis, but without the two endpoints of this axis. It is natural to adjoin these two endpoints. However, these two points lie on the boundary of the unit disk, and that would make any two of these lines intersect in *two* points! What to do? Let's look again at our example of C_1 and C_2. As we rotated one line to become parallel to another, the intersection point flew off in one direction towards infinity. As we rotate beyond parallelism, the intersection point pops up on the other side of the line, moving in the same direction as before. In this respect the two added points act like one ordinary point. To see this more clearly, consider in $\mathbb{R}^2$ a line and a fixed point P on it. Rotate a second line about a point $Q \neq P$. Figure 2.3 depicts the intersection point moving in a uniform direction and passing P.

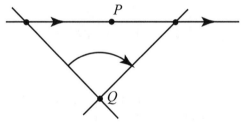

FIGURE 2.3.

The solution to our conundrum: agree that it's really just one point at infinity. In the disk, therefore, identify the two added points and call any line with the adjoined point at infinity the *projective completion* of the line. Doing this for all pairs of antipodal boundary points of the open disk forms a closed disk with each pair of antipodal boundary points identified. This construction is similar to creating a Möbius strip or torus by gluing together, or identifying, appropriate edges of a rectangle. A closed disk with antipodal boundary points identified and supplied with the topology to be given in Definition 2.2 on p. 37 is the *disk model of the real projective plane* $\mathbb{P}^2(\mathbb{R})$.

We'll see that topologically, the projective completion of any line is a loop. Importantly, all the identified antipodal points form a loop, too. Furthermore, this loop intersects any other projectively completed line in a single point at infinity. These facts lead to this definition.

Definition 2.1. The boundary of the disk with antipodal points identified is called the *line at infinity* of $\mathbb{P}^2(\mathbb{R})$.

From this it is not hard to prove our first nice theorem:

Theorem 2.1. Any two different lines of $\mathbb{P}^2(\mathbb{R})$ intersect in exactly one point.

Note that this theorem holds even when one line happens to be the line at infinity.

2.2 Examples

Here are some examples of images of some algebraic curves in our disk model.

Example 2.1. Figure 2.4 depicts the image of the parabola $y^2 = x$ under the shrinking map in (2.1) on p. 31. Begin by parametrizing the parabola by $\{x = t^2, y = t\}$, then divide both x and y by $\sqrt{x^2 + y^2 + 1}$ to obtain the parametrization of the disk image:

$$\left\{ x = \frac{t^2}{\sqrt{t^4 + t^2 + 1}},\ y = \frac{t}{\sqrt{t^4 + t^2 + 1}} \right\}.$$

The added point at infinity completes the image to a closed loop. The

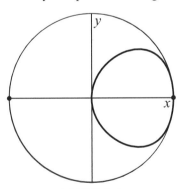

FIGURE 2.4.

picture also shows the point $(-1, 0)$ that is identified to $(1, 0)$. Rotate the curve-plus-point in Figure 2.4 counterclockwise 90° about the origin to get the disk image of $y = x^2$.

Example 2.2. We can apply the same technique to parametrizations of other curves to draw their disk model images. Figure 2.5 shows the disk model of the cusp curve $y^2 = x^3$ parametrized by $\{x = t^2,\ y = t^3\}$, as well as the alpha curve $y^2 = x^2(x+1)$ parametrized by $\{x = t^2 - 1,\ y = t(t^2 - 1)\}$. In each, the antipodal boundary points $(0, 1)$ and $(0, -1)$ of the disk are

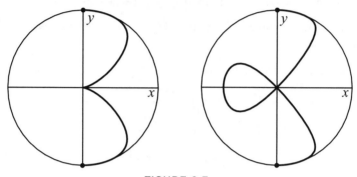

FIGURE 2.5.

identified, making the cusp image a topological loop and the alpha curve image a topological figure 8.

Example 2.3. The phenomenon of the ends closing up in $\mathbb{P}^2(\mathbb{R})$ also holds for polynomial graphs $y = p(x)$. Figure 2.6 shows the image of an odd-degree polynomial.

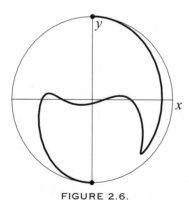

FIGURE 2.6.

The image in $\mathbb{P}^2(\mathbb{R})$ of *any* polynomial $y = p(x)$ with real coefficients is a topological loop.

Example 2.4. The shrinking of $\mathbb{R}^2$ under the map can lead to surprises. Figure 2.7 depicts the images of the four square hyperbolas $x^2 - y^2 = c^2$

2.3. A Basic Picture 35

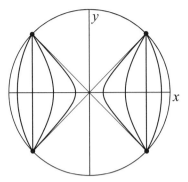

FIGURE 2.7.

for $c^2 = .175, .5, 1, 2$. The nature of the shrinking causes the image of $x^2 - y^2 = 1$ to appear as two vertical line segments. For any $c > 1$ the branches of $x^2 - y^2 = c^2$ bow outward. The two line segments of slope ± 1 are the common asymptotes to the hyperbolas for all c^2. In the disk model, when $c^2 = 1$ or 2, they don't seem to act like asymptotes — that is, "tangent at infinity." Angles are not preserved under the shrinking map, and at infinity the notion of distance itself breaks down, so the metric notion of asymptotic may not look the way we expect. Any hyperbola in the disk $\mathbb{P}^2(\mathbb{R})$ forms a closed loop in the natural topology that we'll introduce in a moment.

2.3 A Basic Picture

The picture in Figure 2.8 accomplishes several things:
- Out of the many ways one could map $\mathbb{R}^2$ to the open unit disk, the picture will motivate our particular choice.
- It will lead to a symmetric definition of $\mathbb{P}^2(\mathbb{R})$.
- It will make it easy to define a natural topology on $\mathbb{P}^2(\mathbb{R})$.
- The picture will provide a way to recenter at infinity, permitting a detailed look at how curves behave there.
- It will lead to a vector space definition of $\mathbb{P}^2(\mathbb{R})$ which will then allow us to generalize easily to $\mathbb{P}^n(\mathbb{R})$ and $\mathbb{P}^n(\mathbb{C})$.

Figure 2.8 shows the unit sphere centered at the origin of $\mathbb{R}^3$ and the plane $z = 1$ parallel to the (x, y)-plane. The picture reveals the geometry behind the shrinking function's formula. Start at any point $(x, y) \in \mathbb{R}^2$, project it vertically to $(x, y, 1)$ in the plane $z = 1$, then radially to the sphere, and then drop down to the (x, y)-plane, landing at $\frac{(x,y)}{\sqrt{x^2+y^2+1}}$,

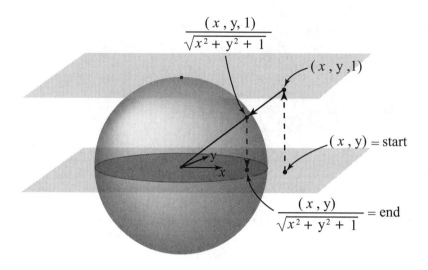

FIGURE 2.8.

which is inside the unit disk. This geometric way of looking at the shrinking map leads to a basic observation. A line in the (x, y)-plane projects up to a line in the plane $z = 1$, which together with $(0, 0, 0) \in \mathbb{R}^3$ defines a plane through $(0, 0, 0)$, and that in turn intersects the upper hemisphere in the top half of a great circle. So the disk image of a typical line is the projection of a great semicircle on the hemisphere. In a plane, the projected image of a circle is an ellipse, fitting in with our earlier observation that the image in the disk of the two lines $y = \pm a$ determines an ellipse.

Vertical projection defines a 1:1-onto map between the disk model and the *hemisphere model* of the real projective plane: the upper hemisphere with opposite equatorial points identified. That's just a step away from looking at $\mathbb{P}^2(\mathbb{R})$ as the entire sphere in which *each* pair of antipodal points is identified to a point. This is the *sphere model* of the real projective plane and is beautifully symmetric, but we can go still further and eliminate the arbitrariness in choosing a particular size of sphere radius. Instead of identifying a point-pair to a point, *identify an entire 1-space to a point*. (A 1-space is any line through the origin.) That is, we regard points of $\mathbb{P}^2(\mathbb{R})$ as 1-subspaces of $\mathbb{R}^3$. In this way, the radius of a sphere centered at the origin becomes irrelevant. $\mathbb{P}^2(\mathbb{R})$ can be looked at as the set of all 1-subspaces of $\mathbb{R}^3$, and we call this the *vector space model* of the real projective plane.

2.4 Basic Definitions

We have now met four different models of the real projective plane. In Definition 2.2, we define a topology on each. For this, recall that a set of basic open sets of $\mathbb{R}^3$ can be taken to be the set of open balls there.

Definition 2.2.
- In the vector space model of $\mathbb{P}^2(\mathbb{R})$, the points are the 1-subspaces of $\mathbb{R}^3$. A typical *basic open set* $\mathcal{O}$ for the natural topology on this model consists of all 1-subspaces of $\mathbb{R}^3$ intersecting any one open ball of $\mathbb{R}^3$.
- The points of the sphere model of $\mathbb{P}^2(\mathbb{R})$ can be taken as the antipodal point pairs of a sphere $\mathcal{S} : x^2 + y^2 + z^2 = 1$ in $\mathbb{R}^3$. A typical *basic open set* for the natural topology on this model consists of all points of $\mathcal{S}$ intersecting any one basic open set $\mathcal{O}$ in the vector space model.
- The points (x, y, z) of the hemisphere model of $\mathbb{P}^2(\mathbb{R})$ can be taken to be those of $\mathcal{S}$ for which $z \geq 0$. A typical *basic open set* for the natural topology on this model is the intersection of the hemisphere with a basic open set of $\mathcal{S}$.
- The points of the disk model of $\mathbb{P}^2(\mathbb{R})$ can be taken to be the projections $(x, y, z) \to (x, y)$ of points in the hemisphere model to points in the disk $x^2 + y^2 \leq 1$. A typical *basic open set* for the natural topology on this model is the projection of a basic open set of the hemisphere.

We constructed $\mathbb{P}^2(\mathbb{R})$ by adjoining points at infinity to the real plane to further our aim of getting a "multiplication theorem." To take advantage of viewing the points of $\mathbb{P}^2(\mathbb{R})$ as the 1-subspaces of $\mathbb{R}^3$, we need to determine what an algebraic curve C in $\mathbb{R}^2$ looks like in this model, and how we then add the points of infinity to the curve. Let's start with a basic example, a line L in $\mathbb{R}^2$.

Example 2.5. Vertically lift a line L in $\mathbb{R}^2$ to a line L' in the plane $z = 1$ in $\mathbb{R}^3$. A point P in L' determines a 1-space of $\mathbb{R}^3$ through P, and therefore a point in the vector space model of $\mathbb{P}^2(\mathbb{R})$. The farther away P is from the origin, the smaller the angle between the 1-space through P and the (x, y)-plane in $\mathbb{R}^3$. For points of L', this angle never quite reaches 0, so the set of 1-subspaces forms the plane through L' and the origin of $\mathbb{R}^3$, minus L. Of course "the angle reaching 0" would mean that the 1-subspace is parallel to the plane $z = 1$, which corresponds to P being at infinity. Adding this line L then gives the entire plane through L' and the origin of $\mathbb{R}^3$. Thus, a projective line in $\mathbb{P}^2(\mathbb{R})$ is represented by a 2-space in $\mathbb{R}^3$. This, together with what we've seen above, shows that in $\mathbb{R}^3$, subspaces of dimension 1, 2 and 3 correspond in $\mathbb{P}^2(\mathbb{R})$ to points, lines and all of $\mathbb{P}^2(\mathbb{R})$. In fact, the 0-dimensional subspace of $\mathbb{R}^3$ — the origin of $\mathbb{R}^3$ — defines the empty set in $\mathbb{P}^2(\mathbb{R})$, to which we assign dimension -1.

To sum up:

Subspaces of dimension n in $\mathbb{R}^3$ correspond to objects of one lower dimension in $\mathbb{P}^2(\mathbb{R})$.

In topological terms, adding L to the union of the 1-subspaces through L' corresponds to taking the topological closure, with L''s point at infinity being the 1-subspace L. We call this closure the *homogenization* of L' in $\mathbb{R}^3$ and denote it by $H(L')$; it's homogeneous in the usual sense: if it contains a nonorigin point Q, then the 1-subspace containing Q is in it, too.

The concept of homogeneous set has an algebraic counterpart. A *homogeneous polynomial* is one in which all terms have the same degree. We can *homogenize* a nonhomogeneous polynomial such as $(x-1)^2 + y^2 - 4$ to get a homogeneous polynomial in x, y, z of degree two. One method: expand the polynomial to $x^2 - 2x + 1 + y^2 - 4$ and then pack each term with whatever power of z is needed to make the polynomial homogeneous, obtaining in this case $x^2 - 2xz + z^2 + y^2 - 4z^2$. An equivalent method: pack the factors directly: $(x-z)^2 + y^2 - 4z^2$. The set-theoretic and algebraic notions of homogeneous are connected through the following.

Theorem 2.2. *The zero set in $\mathbb{R}^n$ of a homogeneous polynomial is homogeneous. If a homogeneous set in $\mathbb{R}^n$ is the zero set of a polynomial, then the polynomial is homogeneous.*

For a slightly more general statement and its proof, see [Kendig 2], Chapter II, Theorem 2.6.

Example 2.6. Let C be a curve defined by a polynomial of degree two. Lifting C in the (x, y)-plane to the plane $z = 1$, passing 1-spaces through each point and then taking the topological closure is a visual way to homogenize C. Doing this to the circle C defined by $(x-1)^2 + y^2 - 4$, for example, produces the cone illustrated in Figure 2.9. The cone is the zero set of the homogenization of $(x-1)^2 + y^2 - 4$. We can rewrite the cone's equation as $x^2 - 2xz + y^2 = 3z^2$, the equation being homogeneous in the obvious sense.

Modeling $\mathbb{P}^2(\mathbb{R})$ either as a sphere with antipodal points identified, or as the 1-subspaces of $\mathbb{R}^3$ has another important consequence: the equator

2.4. Basic Definitions

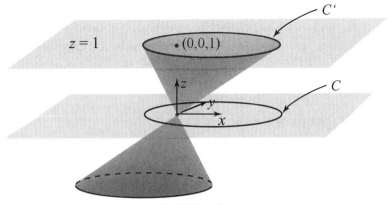

FIGURE 2.9.

no longer plays a special role. This will allow us to recenter anywhere in the projective plane, even at a point at infinity. Recentering at any point P in the projective plane allows us to write the equation of the curve so that P is the new origin, which will let us see in a precise way the behavior of the curve there. This will enable better and more complete bookkeeping and will bring us a step closer to eliminating exceptions and keeping alive the promise of a general multiplication theorem that would count the number of intersection points of two curves.

Example 2.7. We can illustrate this recentering idea using Figure 2.9. The circle C' generates the cone, and we can think of the cone as a global representation of the circle in $\mathbb{P}^2(\mathbb{R})$. We could slice the cone with the affine plane or viewing screen $z = 1$ to get the circle again, but there's no special reason to select $z = 1$ as the screen. We could just as well choose any other plane not passing through the origin of $\mathbb{R}^3$. As we choose various viewing screens this way, we encounter a variety of conic sections — circles, ellipses, parabolas and hyperbolas. In each case we can choose new coordinates $(x, y, z) \in \mathbb{R}^3$ so that the slicing plane has equation $z = 1$. If we then homogenize as before, we end up with the original cone.

Already the $\mathbb{R}^3$ model has shown a remarkable power to unify. All ellipses, parabolas and hyperbolas are simply different views of one and the same projective object. In any viewing screen, the conic section is an *affine curve*. The "global" view, the cone in $\mathbb{R}^3$ thought of as a curve in $\mathbb{P}^2(\mathbb{R})$, is a *projective curve*. Let's make some official definitions.

Definition 2.3. In any real plane with coordinates (x, y), if the zero set of a polynomial $p(x, y)$ defines a curve C in $\mathbb{R}^2$, then C is called a *real affine plane curve* or an *affine curve* in $\mathbb{R}^2$. The homogenization $p(x, y, z)$ of $p(x, y)$ defines a homogeneous zero set in $\mathbb{R}^3$, and the 1-subspaces this set — points in $\mathbb{P}^2(\mathbb{R})$ — comprise a *real projective plane curve* or a *projective curve* in $\mathbb{P}^2(\mathbb{R})$. It is called the *projective completion* of C in $\mathbb{R}^2$.

2.5 Further Examples

In Figure 2.9 the affine viewing screen is $z = 1$ and the points at infinity in it are the 1-subspaces of the plane $z = 0$. From the dimension-lowering observation made on p. 38, this plane, a 2-subspace of $\mathbb{R}^3$, corresponds to a projective line so all points at infinity form a line in $\mathbb{P}^2(\mathbb{R})$. The choice $z = 1$ was arbitrary. Any plane $ax + by + cz = 1$ can be regarded as a viewing plane, and its points at infinity are the lines through the origin in the 2-subspace $ax + by + cz = 0$. For every choice of plane in $\mathbb{R}^3$, we determine a corresponding line at infinity.

We can make our qualitative observations about viewing planes more concrete by using equations.

Example 2.8. For the cone in Figure 2.9, we should be able to write, say, the equation of the hyperbola in which the plane $x = 1$ intersects the cone. Since we know the equation of the cone, this is easy: substitute $x = 1$ into the cone's equation $(x - z)^2 + y^2 = 4z^2$ to get $(1 - z)^2 + y^2 = 4z^2$. This can be rewritten in the standard form

$$\frac{(z + \frac{1}{3})^2}{\left(\frac{2}{3}\right)^2} - \frac{y^2}{\left(\frac{2}{\sqrt{3}}\right)^2} = 1.$$

For a more arbitrary plane parametrized by degree one functions f, g, h as

$$\{x = f(u, v), \ y = g(u, v), \ z = h(u, v)\},$$

substitute $f(u, v)$, $g(u, v)$, $h(u, v)$ in for x, y, z in the cone's equation.

Example 2.9. Everything in the last few paragraphs directly generalizes to any algebraic curve C, giving us a mechanism for obtaining the equation of any algebraic curve in any viewing screen. Coupled with the formula for shrinking the plane to a disk, we can track views of C in the disk model of $\mathbb{P}^2(\mathbb{R})$ as the viewing plane changes. For example, the pictures in Figure 2.10 depict rotated views of the 2×1 ellipse

$$\frac{x^2}{2^2} + \frac{y^2}{1^2} = 1.$$

2.5. FURTHER EXAMPLES 41

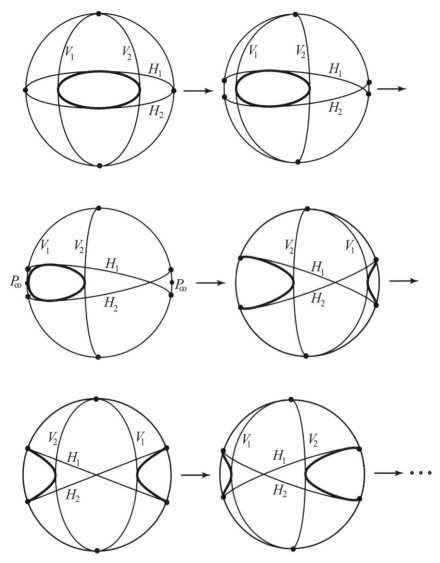

FIGURE 2.10.

Its homogenization $\frac{x^2}{2^2} + \frac{y^2}{1^2} = z^2$ defines an elliptical cone through the origin of $\mathbb{R}^3$, and the original ellipse sits in the plane $z = 1$. The fundamental rectangle surrounding the ellipse extends to the horizontal lines H_1 and H_2 defined by $y = \pm 1$, and the vertical lines V_1 and V_2 given by $x = \mp 2$. Our method allows us to follow the view of the ellipse and these four lines as the viewing plane, always tangent to the unit sphere, slides along a great circle. Figure 2.10 shows six stages of this morphing. These suggest how the sequence continues, finally returning to the original view of the ellipse.

Since changing the viewing screen can take points at infinity and make them finite, we can get equations in any viewing screen, allowing us to analyze what happens around points that originally were at infinity.

Example 2.10. The cubic $y = x^3$ illustrates the power of this method. Its graph is unbounded, and the disk view on the left in Figure 2.11 depicts the curve going through the point Q at the end of the y-axis. If we reposition ourselves at this point, what do we see? To get the answer, form the homogenization $yz^2 = x^3$ and then dehomogenize at the plane $y = 1$ — that is, set $y = 1$. This yields the equation $z^2 = x^3$, which defines a cusp. This figure shows that the axes have switched around; the original line at infinity, which we call the z-axis, has become the new horizontal axis.

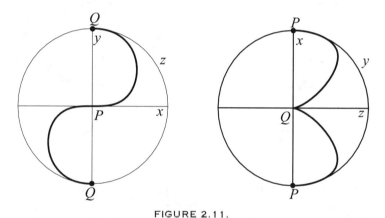

FIGURE 2.11.

Example 2.11. As another example, Figure 2.12 shows views of a rational function's graph — the kind typically encountered when learning to include asymptotes in the graph's sketch.

The rational function is $y = \frac{1}{(x^2-1)(x^2-4)}$, and its graph is an algebraic curve since multiplying the equation by the denominator $(x^2 - 1)(x^2 - 4)$ produces a polynomial equation. The curve has four vertical asymptotes at $x = \pm 1$ and $x = \pm 2$, and the x-axis is a fifth asymptote. The top picture in Figure 2.12 depicts the real affine curve. The disk model of the curve is depicted in the middle picture, and consists of four closed loops — the three obvious ones as well as the left and right branches that connect at the

2.5. Further Examples

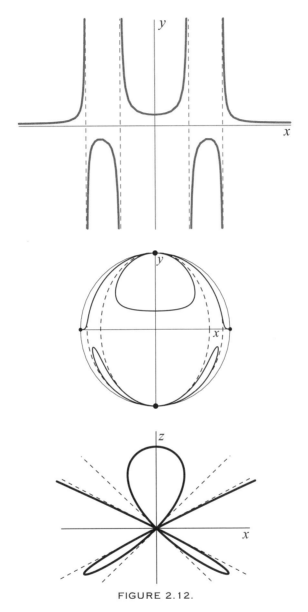

FIGURE 2.12.

ends of the x-axis to make the fourth loop. Notice that the four dashed lines in the original plane appear as two ellipses in the disk model. The curve's polynomial equation $y(x^2 - 1)(x^2 - 4) = 1$ homogenizes to

$$y(x^2 - z^2)(x^2 - 4z^2) = z^5$$

and dehomogenizes at $y = 1$ to

$$(x^2 - z^2)(x^2 - 4z^2) = z^5.$$

Its zero set is depicted in the bottom picture. The point at infinity where the four asymptotes meet has become the new origin, and the four dashed lines represent four lines tangent to this curve at the origin. In this example we see that the phrase "asymptotic means tangent at infinity" is literally true.

CHAPTER 3

From Real to Complex

We have hinted several times at a multiplication theorem for two intersecting algebraic curves, and it is now time to make a promise. In this chapter, we state such a theorem and sketch its proof.

Looking at the parabola $y = x^2$ and the line $y = 1$ suggests what is needed to accomplish our aim. These curves have degree two and one, and intersect in $2 \cdot 1$ points. As we parallel-translate the line downward, the two points of intersection approach each other, and when the line coincides with the x-axis, the points have coalesced, "piling up on each other" at the origin. It is natural to count both intersection points, counting the origin with multiplicity two. When the line is pushed further to $y = -\epsilon$ ($\epsilon > 0$), the curves' intersection points are found from the solutions to $y = x^2$ and $y = -\epsilon$, and these are $x = \pm i \sqrt{\epsilon}$. The two intersection points are therefore $(+i \sqrt{\epsilon}, -\epsilon)$ and $(-i \sqrt{\epsilon}, -\epsilon)$. In the overall downward sweep, the two intersection points begin as real and distinct, approach each other until they meet, then continue as imaginary and distinct. This suggests that working over $\mathbb{C}$ instead of $\mathbb{R}$ allows us to see and keep track of the intersections.

A parabola and line example tells us something else. Look at what happens to the intersection of the parabola $y = x^2$ with the line $y = mx$ as m increases without bound. When m is finite, we see two points of intersection, but when the line becomes vertical, one point has escaped to infinity. This suggests that using the plane $\mathbb{C}^2$ in place of $\mathbb{R}^2$ is still not good enough. It appears that as in the real setting, we need to adjoin points at infinity to $\mathbb{C}^2$ to keep intersection points from leaving our universe. Fortunately our definition of $\mathbb{P}^2(\mathbb{R})$ as the set of all 1-subspaces of $\mathbb{R}^3$ equipped with the topology given in Definition 3.1 generalizes effortlessly.

3.1 DEFINITIONS

Definition 3.1. As a set, $\mathbb{P}^2(\mathbb{C})$ consists of the complex 1-subspaces of $\mathbb{C}^3$. A natural topology is defined by regarding $\mathbb{C}^3$ as $\mathbb{R}^6$ with open 6-balls as a basis for the open sets of $\mathbb{C}^3$. *A basic open set of* $\mathbb{P}^2(\mathbb{C})$ consists of the set of all complex 1-spaces of $\mathbb{C}^3$ intersecting some one basic open set of $\mathbb{C}^3$.

Notice that Definition 3.1 is analogous to the first part of Definition 2.2 on p. 37. The next definition is analogous to Definition 2.3.

Definition 3.2. Let $\mathbb{C}^2$ have complex coordinates (x, y). An *affine curve in* $\mathbb{C}^2$ is the zero set in $\mathbb{C}^2$ of a nonconstant polynomial $p(x, y)$. The homogenization $p(x, y, z)$ of $p(x, y)$ defines a homogeneous zero set in $\mathbb{C}^3$ consisting of complex 1-spaces. The complex 1-spaces of this set, regarded as points in $\mathbb{P}^2(\mathbb{C})$, comprise a *complex projective plane curve* or a *projective curve* in $\mathbb{P}^2(\mathbb{C})$. The projective curve defined by the homogenization of $p(x, y)$ is the *projective completion* of the affine curve $C(p(x, y))$. We denote the affine curve by $C(p)$, and if no confusion can arise, we denote its projective completion by $C(p)$, too. $p(x, y)$ is irreducible in $\mathbb{C}[x, y]$ if and only if its homogenization $p(x, y, z)$ is irreducible in $\mathbb{C}[x, y, z]$. In that case, the affine and projective curves are called *irreducible*.

Comment 3.1. It can be shown that the projective completion of a complex affine curve coincides with the topological closure of the affine curve in $\mathbb{P}^2(\mathbb{C})$. The real analog of this is not true because a real curve may have isolated points at infinity. The topological closure of the real affine curve would not include them, and the closure wouldn't be a real projective algebraic curve. To get an example of this, look at the sketch in the second row, second column of Figure 1.3 on p. 8, showing one isolated point. The equation $y^2 = x^2(x-1)$ puts this isolated point at the origin, and homogenizing and dehomogenizing this at $x = 1$ yields the rational function $z = \frac{1}{1+y^2}$ whose graph in $\mathbb{R}^2$ is an algebraic curve, but whose topological closure in $\mathbb{P}^2(\mathbb{R})$ is not an algebraic curve there.

With the above definitions, we are now ready to address in earnest our goal of a multiplication theorem. Suppose C_1 and C_2 are affine curves of degree m and n, and that in $\mathbb{R}^2$ they intersect in mn distinct points. As we move the curves around or modify their equation coefficients without changing their degrees, there are three ways the mn points can escape good bookkeeping:

- They can escape from $\mathbb{R}^2$ into $\mathbb{C}^2$.
- They can escape to infinity.
- They can pile up on each other, forming points with multiplicity.

By choosing our universe to be $\mathbb{P}^2(\mathbb{C})$, we'll be able to deal successfully with the first two of these. It remains to assign a multiplicity to each point of intersection. We do this next.

3.2 The Idea of Multiplicity; Examples

We often first hear about multiplicity in a version of the Fundamental Theorem of Algebra. One form of it says that any nonconstant monic polynomial $p(x) \in \mathbb{C}[x]$ factors into a product of $\deg(p)$ monic linear factors, $p(x) = \prod (x-a_i)^{m_i}$ ($a_i \in \mathbb{C}$), unique up to the order of factors. The exponent m_i is called the *multiplicity of* a_i and is the total number of occurences of $(x - a_i)$ in the factorization. This algebraic notion is often supplemented by the geometric picture of a total of $\deg(p)$ points a_i in $\mathbb{C}$, with m_i points piled up at each a_i.

The Fundamental Theorem of Algebra can be translated into a statement about two plane algebraic curves in $\mathbb{C}^2$, revealing its close connection to our subject. One curve is the graph in $\mathbb{C}^2$ of $y = p(x)$ and the other is the complex line $y = 0$. Consider this example:

Example 3.1. Let $p(x) = x^2$. Then $p(x) = \prod(x - a_i)^{m_i} = (x - 0)^2$, so the only zero is $a = 0$, and it occurs with multiplicity 2. The corresponding two-curve picture consists of the parabola $y = x^2$ and the x-axis. In $\mathbb{R}^2$, we can see this multiplicity in a more dynamic way, and it suggests a natural way to generalize intersection multiplicity to any two plane algebraic curves. To introduce the idea, look at $y = 0$ as the limit of $y = \epsilon$ ($\epsilon > 0$) as $\epsilon \to 0$. The intersection of the parabola with $y = \epsilon$ consists of the two points $(\pm \sqrt{\epsilon}, \epsilon)$, and as $\epsilon \to 0$, they approach each other, coalescing at the origin. Geometrically, we've taken the x-axis, translated it upward a bit and then let it float back down to its original position. We could instead push the line $y = -\epsilon$ upward: there are two intersection points $(\pm i \sqrt{\epsilon}, -\epsilon)$ that in the limit coalesce at the origin. In either case, we've perturbed the x-axis slightly to separate the two "stuck-together" points where the parabola and the x-axis intersect. This suggests that the multiplicity of this intersection is the number of single points that flow together as the x-axis returns to its original position.

This simple example, though it's in the real setting, nevertheless captures the heart of how we'll arrive at a definition of intersection multiplicity. In the next example we once again slightly perturb one curve to separate stuck-together points.

Example 3.2. The two cusps $y^2 = x^3$ and $y^3 = x^2$ intersect at the origin. What is the multiplicity of this intersection? We can translate either cusp to get the picture in Figure 3.1.

FIGURE 3.1.

Near the origin we see two cusps intersecting in four points. As in the parabola example, there's a translation separating all stuck-together points to make them individually visible in $\mathbb{R}^2$. As in the parabola example, other translations can yield intersection points with imaginary components, where we see only two or none in $\mathbb{R}^2$. What's more, ϵ could be a small complex number. If we could easily envision $\mathbb{C}^2$ with its four real dimensions, this approach might offer a workable approach to counting points. But for us three-dimensioneers, we need to supplement it with algebra for reliable results.

We begin the algebraic approach by looking at m_i in the factorization $p(x) = \prod (x - a_i)^{m_i}$ in a slightly different light, as the order of p at a_i.

Definition 3.3. The *order* of a polynomial $p(x)$ at a is the degree of the initial term of p when expanded about a.

This extends Definition 1.2, on p. 23, and says that if

$$p(x) = c_0(x - a)^m + c_1(x - a)^{m+1} + c_2(x - a)^{m+2} + \cdots \quad (c_0 \neq 0),$$

then the order of p at a is m. To show that m is in fact the multiplicity of the root a in the Fundamental Theorem, let's simplify notation by assuming coordinates have been chosen so that $a = a_1 = 0$. The factorization then reads $p(x) = x^{m_1} \prod_{i>1}(x - a_i)^{m_i}$, where for all $i > 1$, $a_i \neq 0$. Therefore the expansion of $\prod_{i>1}(x - a_i)^{m_i}$ has a nonzero constant term c equal to

3.2. THE IDEA OF MULTIPLICITY; EXAMPLES

$\prod_{i>1}(-a_i)^{m_i}$. The lowest-degree term of $x^{m_1} \prod_{i>1}(x-a_i)^{m_i}$ is then cx^{m_1} ($c \in \mathbb{C} \setminus \{0\}$), so the order of $p(x)$ at any point a is the multiplicity of a as a root.

When x is very small, $\prod_{i>1}(x-a_i)^{m_i}$ stays close to c, so $p(x)$ remains close to cx^m. We look at the root 0 of p as the limit of intersections of the graphs of $y = p(x)$ and $y = \epsilon$ as $\epsilon \to 0$. From what we've just said, cx^m approximates p around 0. For our purposes we can take $c = 1$, since the geometric effect of multiplying by c is expansion or contraction together with rotation about the origin, none of which changes any relevant behavior. Therefore we can calculate the multiplicity of the root 0 of p by looking at the limit of the number of intersections of the graphs of $y = x^m$ and $y = \epsilon$ as $\epsilon \to 0$.

This is easy enough. To find the intersection of the graphs of $y = x^m$ and $y = \epsilon$, substitute $y = \epsilon$ into $y = x^m$, thus getting the equation $\epsilon = x^m$. This has m solutions, $\epsilon^{1/m}$ times the mth roots of unity, and we may look at these m points as residing in the complex line $y = \epsilon$. As ϵ goes from small positive to zero, the m intersections follow rays toward the origin enclosing angles of $\frac{2\pi}{m}$. As ϵ goes from zero to small negative, the intersections follow rays away from the origin along the angle bisectors. Figure 3.2 illustrates the idea for $m = 3$.

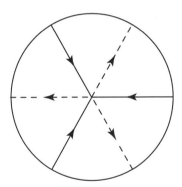

FIGURE 3.2.

Importantly, ϵ can be taken to be complex in the above. As $\epsilon \to 0$ by multiplying by ever smaller real scalars, Figure 3.2 changes by being rotated. Continuously approaching 0 in fancier ways leads to fancier alternating "in" and "out" paths, but the basic spirit of the figure remains the same. These considerations suggest an algebraic way to define intersection multiplicity. Let C_1 and C_2 be two curves intersecting at the origin. Suppose that C_1 is the zero set of $F(x, y)$, and that C_2 has a 1:1 parametrization by polynomials, $\{x = p(t), y = q(t)\}$, with the origin corresponding to $t = 0$. (A parametrization by polynomials is special, but right now we're just motivating things.) If C_1 is the x-axis and C_2 is the parabola considered above, we can use $F(x, y) = y$ and the parametrization $\{x = t, y = t^2\}$. Or, in the example of the two cusps, $F(x, y)$ can be $y^2 - x^3$ and the parametrization, $\{x = t^3, y = t^2\}$. As t fills out a small neighborhood of 0 in $\mathbb{C}$, the parametrization fills out the part of C_2 within some neighborhood of $(0, 0)$ in $\mathbb{C}^2$. This leads to a very consequential question:

What can be said about $F(p(t), q(t))$?

The argument $(p(t), q(t))$ of F is constrained to lie on C_2. Therefore $F(p(t), q(t))$ is zero only when the point $(p(t), q(t))$ is on both C_1 and C_2. We know $(0, 0) \in \mathbb{C}^2$ is such a point, and it corresponds to $t = 0$. Therefore the order of t in the polynomial $F(p(t), q(t))$ is some positive r.

Now let's apply our "perturbation" philosophy by replacing $F(x, y)$ by $F(x, y) - \epsilon$. This changes the curve C_1 to some $\tilde{C}_1$. Intuitively, ϵ can be chosen so small that within any neighborhood of $(0, 0) \in \mathbb{C}^2$ our eyes can't see any difference between C_1 and $\tilde{C}_1$. The ϵ simply changes F's constant term, so it changes the polynomial $F(p(t), q(t))$ to $F(p(t), q(t)) - \epsilon$. The zero set of this can be looked at as the intersection of the graphs of $y = F(p(t), q(t))$ and $y = \epsilon$. But we met this just a moment ago! This intersection in $\mathbb{C}$ consists of $r \geq 1$ points coalescing to $0 \in \mathbb{C}$, the situation illustrated in Figure 3.2. So adding a small quantity to F perturbs C_1, separating any coalesced points, and the order of $F(p(t), q(t))$ tells us just how many points on $C_1 \cap C_2$ have coalesced at $(0, 0) \in \mathbb{C}^2$.

Let's see how this works for the two examples of the parabola-and-line and the two cusps.

Example 3.3. Let the curves be the x-axis and parabola $y = x^2$. Then $F(x, y) = y$ defines the x-axis and a parametrization of the parabola is $\{x = t, y = t^2\}$. Then $F(p(t), q(t)) = t^2$ which has order 2. The intersection multiplicity of the line and parabola is therefore 2.

Example 3.4. In the example of two cusps, let $F(x, y) = y^2 - x^3$ define one cusp and let $\{x = t^3, y = t^2\}$ be a parametrization of the other. Then $F(p(t), q(t)) = (t^2)^2 - (t^3)^3 = t^4 - t^9$, which has order 4. Thus the perturbation depicted in Figure 3.1 does in fact separate all points at the origin.

3.2. The Idea of Multiplicity; Examples

Example 3.5. We can apply our method to a more subtle example in which looking at perturbations in $\mathbb{R}^2$ is of little help. Let C_1 be the cusp $y^2 - x^3$ and C_2, the cusp $4y^2 - x^3$, shown on the left side of Figure 3.3. The x-axis

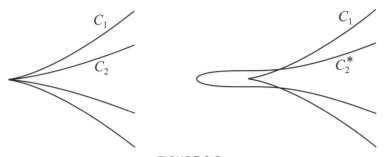

FIGURE 3.3.

is tangent to both curves at the origin because the line through $(0, 0)$ and a point $P \neq (0, 0)$ on either curve approaches the x-axis as P approaches the origin along that curve. Therefore the two cusps are in a natural sense mutually tangent at the origin, and such a "higher order of contact" usually increases intersection multiplicity. But in $\mathbb{R}^2$, translating either cusp yields at most two points of intersection. For example, the picture on the right in Figure 3.3 shows the effect on C_2 of subtracting $\epsilon = 0.001$ from $4y^2 - x^3$. Any such perturbation produces at most two separated points near the origin in $\mathbb{R}^2$. It is in $\mathbb{C}^2$ that the higher multiplicity due to tangency reveals itself. These perturbations split the cusps' intersection into *six* separate points in $\mathbb{C}^2$. The algebraic approach reflects this: C_1 is defined by $F = y^2 - x^3$ and a parametrization of $C_2 = C(4y^2 - x^3)$ is $\{x = t^2, y = \frac{t^3}{2}\}$. Therefore $F(p(t), q(t)) = \left(\frac{t^3}{2}\right)^2 - (t^2)^3 = -\frac{3}{4}t^6$, which has order 6.

In finding the intersection multiplicity in Examples 3.3 through 3.5 we always chose to parametrize C_2. It turns out we could equally well have chosen to parametrize C_1. This essentially says that multiplicity is well-defined. Let's check this in each case.

Example 3.6. For the x-axis and parabola $y = x^2$, let $F(x, y) = y - x^2$ and parametrize the line by $\{x = t, y = 0\}$. Then $F(t, 0) = t^2$. At $(0, 0)$ this has order 2 and therefore the intersection multiplicity is 2 there.

Example 3.7. For the two cusps in Example 3.4, let $F(x, y) = y^3 - x^2$ and parametrize the other cusp by $\{x = t^2, y = t^3\}$. Then $F(t^2, t^3) = t^9 - t^4$, so the cusps' intersection multiplicity at the origin is again 4.

Example 3.8. For the two tangent cusps in Example 3.5, let $F(x, y)$ be $4y^2 - x^3$ and parametrize the other cusp by $\{x = t^2, y = t^3\}$. Then $F(t^2, t^3) = 4t^6 - t^6 = 3t^6$, so the two tangent cusps' intersection multiplicity at the origin is 6.

3.3 A REALITY CHECK

We assumed in the last section that within some neighborhood of the origin, one of the curves has a 1:1 parametrization $\{x = p(t), y = q(t)\}$ with p, q polynomials. This assumption is quite special. For one thing, it may take more than one parametrization to describe the curve. For example, the curve $y^2 - x^2 = 0$ defines two lines through the origin, and each requires a parametrization: $\{x = t, y = t\}$ for the line of slope 1, and $\{x = t, y = -t\}$ for the line of slope -1. Also, it may not be possible to choose p and q to be polynomials.

Example 3.9. The real part of the alpha curve $y^2 = x^2(x + 1)$ near the origin of $\mathbb{C}^2$ is depicted in Figure 3.4.

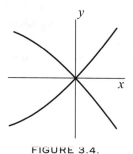

FIGURE 3.4.

Solving for y gives $y = x\sqrt{x+1}$ and $y = -x\sqrt{x+1}$. Expanding $x\sqrt{x+1}$ in a Maclaurin series gives

$$f(x) = x + \frac{1}{2}x^2 - \frac{1}{8}x^3 + \frac{1}{16}x^4 - \frac{5}{128}x^5 + \frac{7}{256} - \cdots,$$

valid throughout some open disk in $\mathbb{C}$ centered at $x = 0$. The part in Figure 3.4 is then given by the two parametrizations $\{x = t, y = \pm f(t)\}$. The part filled out by each parametrization as t fills out the disk is called an *analytic branch* of the curve through the origin. The alpha curve therefore has two analytic branches through $(0, 0)$. If P is any point of a curve, we may translate coordinates to make P the origin, so we may speak of analytic branches of a curve through any of its points. We often call an analytic branch simply a *branch*.

Example 3.10. What happens if we change the alpha curve's equation $y^2 = x^2(x+1)$ to $y^2 = x(x+1)$? This defines a hyperbola, and since $y = \pm x^{\frac{1}{2}}\sqrt{x+1}$, its right branch is tangent to the y-axis at the origin. The Maclaurin series for $y = \sqrt{x+1}$ multiplied by $x^{\frac{1}{2}}$ yields the "fractional-power series"

$$x^{\frac{1}{2}} + \frac{1}{2}x^{\frac{3}{2}} - \frac{1}{8}x^{\frac{5}{2}} + \frac{1}{16}x^{\frac{7}{2}} - \cdots.$$

It leads to a parametrization just as easily as an ordinary power series does: simply set $t = x^{\frac{1}{2}}$ to obtain a parametrization about the origin:

$$\left\{ x = t^2, \ y = t + \frac{1}{2}t^3 - \frac{1}{8}t^5 + \frac{1}{16}t^7 - \cdots \right\}.$$

Altering $y^2 = x(x+1)$ to, say, $y^6 = x^{11}(x+1)^3$ leads in a similar way to a fractional power series

$$x^{\frac{11}{6}} + \frac{1}{2}x^{\frac{17}{6}} - \frac{1}{8}x^{\frac{23}{6}} + \frac{1}{16}x^{\frac{29}{6}} - \cdots,$$

and setting $t = x^{\frac{1}{6}}$ yields the parametrization

$$\left\{ x = t^6, \ y = t^{11} + \frac{1}{2}t^{17} - \frac{1}{8}t^{23} + \frac{1}{16}t^{29} - \cdots \right\}.$$

We arrived at the above parametrizations by solving for y, and the solutions actually amount to a factorization. That is, $y^2 - x(x+1)$ factors into the product $\left(y - \sqrt{x(x+1)}\right) \cdot \left(y + \sqrt{x(x+1)}\right)$, and within a sufficiently small disk this factorization can be written as

$$\left[y - \left(x^{\frac{1}{2}} + \frac{1}{2}x^{\frac{3}{2}} - \frac{1}{8}x^{\frac{5}{2}} + \cdots \right) \right] \left[y + \left(x^{\frac{1}{2}} + \frac{1}{2}x^{\frac{3}{2}} - \frac{1}{8}x^{\frac{5}{2}} + \cdots \right) \right].$$

Remarkably, factoring polynomials using fractional power series in this way holds more generally, and since each factor can be converted to a parametrization, factoring is the key to analyzing the local behavior of plane algebraic curves in $\mathbb{C}$. We now turn to this central result.

3.4 A Factorization Theorem for Polynomials in $\mathbb{C}[x, y]$

We begin with a few assumptions. Let $p(x, y)$ be a polynomial in $\mathbb{C}[x, y]$ of degree $n \geq 1$ in y. Choose coordinates so that p has the form

$$y^n + p_1(x)y^{n-1} + \cdots + p_n(x)$$

with $p_i \in \mathbb{C}[x]$. We can always do this, because if p isn't already of this form, then apply a linear shear sending x to $x + \epsilon y$, with $\epsilon \in \mathbb{R}$ nonzero and small. This makes the coefficient of y^n a nonzero polynomial in ϵ, say $\tilde{p}(\epsilon)$. Choose ϵ_0 so that $\tilde{p}(\epsilon_0) \neq 0$, and then divide p by $\tilde{p}(\epsilon_0)$.

Comment 3.2. For future reference, we note that ϵ_0 may be chosen to work simultaneously for two polynomials p, q of degree m and n: under $x + \epsilon y$, the coefficients of y^m and y^n are polynomials in ϵ, say $\tilde{p}(\epsilon)$ and $\tilde{q}(\epsilon)$. Choose ϵ_0 so that $\tilde{p}(\epsilon_0)$ and $\tilde{q}(\epsilon_0)$ are both nonzero, then divide p by $\tilde{p}(\epsilon_0)$ and q by $\tilde{q}(\epsilon_0)$.

Let f_i denote a function that is complex-analytic in a neighborhood of $0 \in \mathbb{C}$. Let $x^{\frac{1}{r}}$ be a symbol satisfying $(x^{\frac{1}{r}})^s = x^{\frac{s}{r}}$, with $x^{\frac{1}{1}} = x$, and assume these symbols have the expected algebraic properties. We state without proof this factorization theorem.

Theorem 3.1. There exists a unique set of f_i and associated positive integers r_i such that for all x in some neighborhood of $0 \in \mathbb{C}$,

$$y^n + p_1(x)y^{n-1} + \cdots + p_n(x) = \prod_{i=1}^{n} \left(y - f_i(x^{\frac{1}{r_i}}) \right). \quad (3.1)$$

Comment 3.3. There are two principal approaches to proving this theorem. One way establishes the existence of the factorization, but doesn't provide a method of constructing the factors. An example of this is found in Chapter 13 of [Picard], Vol II; the proof uses complex-analytic arguments. Another approach, considerably longer and more involved, provides an algorithm for producing the factors. A proof of this type is found in Chapter IV of [Walker] and uses the Newton polygon. This polygon is constructed by plotting i versus the order of $p_i(x)$ at $x = 0$. This defines a set of lattice points in the first quadrant of $\mathbb{Z} \times \mathbb{Z}$, and we form the set's convex hull, the boundary of which is the Newton polygon. Information about the sides of this polygon leads to values for r_i. Once these are found, the coefficients c_i in $\sum_i c_i x^{\frac{1}{r_i}}$ can be determined using substitution. This can be time-consuming, but there are routines in packages like Maple and Mathematica that automatically compute the fractional-power series to any desired degree of accuracy.

3.5 LOCAL PARAMETRIZATIONS OF A PLANE ALGEBRAIC CURVE

Let $p(x, y)$ define an algebraic curve C in $\mathbb{C}^2$, and let P be any point of C. Suppose coordinates in $\mathbb{C}^2$ have been chosen so that P is the origin and so that $p(x, y)$ has the form $y^n + p_1(x) y^{n-1} + \cdots + p_n(x)$. By (3.1) we can write

$$p(x, y) = \prod_{i=1}^{n} \left(y - f_i(x^{\frac{1}{r_i}}) \right).$$

Since $(0, 0) \in C$, at least one of the f_i must have positive order at 0 — that is, the lowest power of $x^{\frac{1}{r_i}}$ in f_i is positive. Assume indices have been chosen so that $f_1, \ldots, f_s$ ($0 < s \le n$) are those f_i with positive order at 0. A parametrization $(t^{r_i}, f_i(t))$ is associated with each factor, and these define all the branches of C through $(0, 0) \in \mathbb{C}^2$.

For the alpha curve in Example 3.9 on p. 52, we obtained two parametrizations and two associated branches through the origin. In each case the fractional power series in the factorization is an ordinary integral power series. The next example tells an important story.

Example 3.11. The polynomial $p(x, y) = y^2 - x^3$ defining a cusp curve factors into $(y - x^{\frac{3}{2}}) \cdot (y + x^{\frac{3}{2}})$. Here $r_1 = r_2 = 2$. Each of $f_1 = x^{\frac{3}{2}}$ and $f_2 = -x^{\frac{3}{2}}$ has positive order at 0, so each defines a branch of the cusp curve through the origin. But there is actually only *one* branch of the curve through the origin! So although the factor $(y - x^{\frac{3}{2}})$ defines the parametrization $\{x = t^2, y = t^3\}$ and $(y + x^{\frac{3}{2}})$ defines a different parametrization $\{x = t^2, y = -t^3\}$, the same branch is filled out by each as t fills out a neighborhood of $\mathbb{C}$ about 0. Now $x^{\frac{1}{2}}$ is determined only up to ± 1 — that is, up to a second root of unity. More generally, $x^{\frac{1}{r_i}}$ is determined only up to an r_ith root of unity. The above factors group themselves in a natural way into disjoint classes, each containing r_i mutually conjugate roots, with all conjugate roots defining the same branch. Therefore, in our factorization we will encounter r_i different fractional-power series, one in each of the variables

$$x^{\frac{1}{r_i}}, \; \epsilon x^{\frac{1}{r_i}}, \; \epsilon^2 x^{\frac{1}{r_i}}, \; \ldots, \; \epsilon^{r_i-1} x^{\frac{1}{r_i}},$$

where ϵ is a primitive r_ith root of unity. Setting $t = x^{\frac{1}{r_i}}$, we obtain a corresponding parametrization $\{x = t^{r_i}, \; y = f(t)\}$. The first part $x = t^{r_i}$ is the same no matter which of the conjugates we choose. The choice of conjugate *does* affect the parametrization of y, but within some neighborhood of $(0,0) \in \mathbb{C}^2$, all r_i mutually associated parametrizations fill out the same branch of the curve C.

3.6 Definition of Intersection Multiplicity for Two Branches

Theorem 3.2. Let P be an isolated point in the intersection of two curves C_1 and C_2 in $\mathbb{C}^2$. Suppose coordinates in $\mathbb{C}^2$ have been chosen so that P is the origin, and suppose C_1 and C_2 are defined in these coordinates by square-free polynomials p and q. Let the part of C_1 near the origin be a single analytic branch, and the same for C_2, so that these parts have respective parametrizations $\{x = t^{r_i}, \; y = f_i(t)\}$ and $\{x = t^{s_j}, \; y = g_j(t)\}$. Denote the order of a power series in t by o. Then

$$\sum_j o\big(p(t^{s_j}, g_j(t))\big) = \sum_i o\big(q(t^{r_i}, f_i(t))\big). \tag{3.2}$$

For a proof of this theorem see [Walker] Theorem 5.1, p. 109–110. We make the following definition.

Definition 3.4. The common value in Theorem 3.2 is called the *intersection multiplicity of C_1 and C_2 at the origin*.

Some intuition. Suppose the origin is a point of intersection of irreducible curves C_1 and C_2 and let's suppose we can see in four dimensions. Let S be a small sphere $x_1^2 + x_2^2 + y_1^2 + y_1^2 = \epsilon^2$ centered at the origin in $\mathbb{R}^4$. For ϵ small enough, the sphere intersects C_1 and C_2 in two disjoint real loops (closed curves), and these loops have a mutual linking number, the number of times one curve winds around the other. This is a purely homotopic concept, so keeping the loops disjoint, continuously deform C_1 so it becomes a circle. The linking number is easily visualized as the number of times C_2 winds around that circle. The choice of making C_1 the circle is arbitrary, and continuously reshaping C_2 so *it* ends up as the circle then causes C_1 to wind around C_2. The geometrical content of Equation (3.2) is this: *These linking* *numbers are the same, and is the intersection multiplicity of C_1 and C_2 at the origin.* It is instructive to physically experiment with this using string. Also, the command **plot_knots** in Maple's package **with(algcurves);** puts slender-tube versions of both closed curves on the screen, and moving the mouse easily changes their orientation, mimicking a physical model.

3.7 An Example

In this section we give an example in which geometric intuition in $\mathbb{R}^2$ is of little use. Instead, it showcases the power of the algebraic approach.

Example 3.12. The part of $C_1 = C\left(x^6 - x^2y^3 - y^5\right)$ in $\mathbb{R}^2$ appears in Figure 3.5 as the more heavily drawn curve. It looks everywhere smooth,

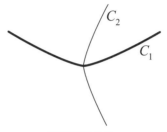

FIGURE 3.5.

but in $\mathbb{C}^2 = \mathbb{R}^4$ we'd see three different branches (think of small disks) of C_1 passing through the origin. Topologically, in some neighborhood of the origin, we'd see the three disks touching in just the origin. In $\mathbb{R}^2$ only one of these branches appears as a curve. Another disk is tangent to the complex line $y = ix$, and the third is tangent to $y = -ix$. Both tangent lines intersect the real plane in only the origin, and so do the associated branches. That's why looking at C_1 is so misleading. Interchanging x and y in $p(x, y)$ to get $p(y, x)$ reflects the real curve C_1 about the line $y = x$, producing the curve C_2. It too is smooth in $\mathbb{R}^2$, but not in $\mathbb{C}^2$.

Because the curves resemble crossing parabolas, to the unsuspecting they appear to intersect with multiplicity 1. However, open Maple's algebraic curves library using the command **with(algcurves);** and then invoke the **puiseux** command. This produces these parametrizations of the three branches of C_1 through the origin of $\mathbb{C}^2$:

$$\left\{x = t, \ y = it + \frac{1}{2}t^2 + \frac{7i}{8}t^3 - \frac{5}{2}t^4 + \cdots\right\},$$

$$\left\{x = t, \quad y = -it + \frac{1}{2}t^2 - \frac{7i}{8}t^3 - \frac{5}{2}t^4 + \cdots\right\},$$

$$\left\{x = t^3, \quad y = t^4 - \frac{1}{3}t^6 + \frac{4}{9}t^8 - \frac{65}{81}t^{10} + \cdots\right\}. \tag{3.3}$$

In the first two parametrizations, neglecting terms of order greater than 1 yields parametrizations of the complex lines $y = \pm ix$, each tangent to C_1 at the origin. There are actually two other parametrizations conjugate to the bottom parametrization (3.3), obtained by everywhere replacing t by t times a third root of unity — that is, by $t\left(-\frac{1}{2} + \frac{i\sqrt{3}}{2}\right)$ or by $t\left(-\frac{1}{2} - \frac{i\sqrt{3}}{2}\right)$. This doesn't affect the parametrization for x, but it does for y. The same branch is filled out for any of the third roots of unity.

Successively substituting the three branch parametrizations of C_1 into C_2's defining polynomial $p(y, x) = y^6 - y^2x^3 - x^5$ yields orders 6, 6, and 15 at the origin, giving a total intersection multiplicity there of $6 + 6 + 15 = 27$. For almost all complex perturbations of the coefficients of $p(x, y)$ and $p(y, x)$ defining perturbed curves $\tilde{C}_1$ and $\tilde{C}_2$, we'd see near $(0, 0) \in \mathbb{C}^2$ the following scene: 27 disks in $\tilde{C}_1$, each about a distinct intersection point of $\tilde{C}_1 \cap \tilde{C}_2$ near $(0, 0)$. Each of these 27 disks can be made sufficiently small so that they're mutually disjoint. We'd also see another 27 such disks in $\tilde{C}_2$ about those same 27 intersection points of $\tilde{C}_1 \cap \tilde{C}_2$. These 54 disks intersect in pairs, each of the 27 disk-pairs touching at a different one of the 27 intersection points separated by the perturbation.

3.8 Multiplicity at an Intersection Point of Two Plane Algebraic Curves

Definition 3.4 applies to two intersecting analytic branches, but it is easy to extend this definition to an intersection point of any two plane algebraic curves. The basic idea is illustrated by this example: let the curve C_1 be m randomly-selected lines through the origin, and C_2, n randomly-selected lines through the origin. Due to the randomness, each line of C_1 intersects each line of C_2 in one point, so there are mn points of $C_1 \cap C_2$ piled up at the origin. The following definition generalizes this.

Definition 3.5. Let P be an isolated point in the intersection of two curves C_1 and C_2 in $\mathbb{C}^2$. Suppose coordinates in $\mathbb{C}^2$ have been chosen so that P is the origin, and suppose C_1 and C_2 are defined in these coordinates by square-free polynomials p and q. Let the part of C_1 near the origin consist of branches $B_{1,1}, \ldots, B_{1,\rho}$, and the part of C_2 near the origin consist of branches $B_{2,1}, \ldots, B_{2,\sigma}$. If $B_{1,i}$ intersects $B_{2,j}$ with multiplicity $m_{i,j}$, then C_1 and C_2 intersect at the origin in multiplicity $\sum_{1,1}^{\rho,\sigma} m_{i,j}$.

3.9 INTERSECTION MULTIPLICITY WITHOUT PARAMETRIZATIONS

In geometrically defining the multiplicity of intersection of two curves at a point P, the idea has been to slightly alter coefficients of one or both defining polynomials in such a way that P splits up into as many separate points as possible, and then count the separated points. As Example 3.12 shows, our limitations in four-dimensional visualization can make this approach misleading and unreliable. The algebraic approach frees us from such visual limitations, but can require considerable computation since it involves breaking up one curve into parametrized branches through P and substituting one from each conjugacy class into the polynomial defining the other curve. The sum of the resulting orders for each substitution is then the multiplicity of intersection at P.

To save work, it would be nice indeed if we could avoid splitting the curve into branches through P, and somehow simply substitute one polynomial directly into another, as we did when one curve is a line. It turns out that the resultant does essentially this, thereby providing an elegant and efficient way of handling intersection multiplicities. The key to understanding how and why resultants can accomplish this seeming miracle comes from an alternative way of defining them. We begin with some motivating examples.

Example 3.13. Let C_1 and C_2 be the graphs of the polynomial functions $y = u_1(x)$ and $y = v_1(x)$. The x-values for which the graphs intersect are given by $u_1(x) = v_1(x)$, so these x-values form the zero set of $u_1(x) - v_1(x)$. If we add more to the curve C_1 by taking its union with the graph of another polynomial function $y = u_2(x)$, then the set of x-values for which the new, larger C_1 intersects C_2 is just the union of the zero sets of $u_1(x) - v_1(x)$ and $u_2(x) - v_1(x)$. This, in turn, is the zero set of the product $(u_1(x) - v_1(x)) \cdot (u_2(x) - v_1(x))$. In a way similar to this, we see that if $C_1 = \cup_{i=1}^m C(y - u_i(x))$ and $C_2 = \cup_{j=1}^n C(y - v_j(x))$, then the set of x-coordinates of the points where C_1 and C_2 intersect is the zero set of

the product
$$\prod_{i,j=1}^{m,n} (u_i(x) - v_j(x)).$$

Starting on p. 13, we explored some properties of the resultant. Impressive though it was then, we are about to reveal its even greater powers.

Example 3.14. Consider again $C_1 = C(y - u_1(x))$ and $C_2 = C(y - v_1(x))$ in Example 3.13. To eliminate y, take
$$\text{resultant}(y^1 - u_1(x)y^0, y^1 - v_1(x)y^0, y)$$
which is
$$\begin{vmatrix} 1 & -u_1(x) \\ 1 & -v_1(x) \end{vmatrix} = u_1(x) - v_1(x).$$
In this case, the resultant computes the difference, whose zeros are the x-values above which C_1 and C_2 intersect. This isn't too impressive just yet, but try this on the larger version
$$C_1 = C((y - u_1(x)) \cdot (y - u_2(x))).$$
This curve is the zero set of $(y - u_1(x)) \cdot (y - u_2(x))$, and this product expands to
$$1 \cdot y^2 - (u_1(x) + u_2(x)) \cdot y^1 + u_1(x) \cdot u_2(x) \cdot y^0.$$
The resultant with respect to y of this and the polynomial $y - v_1(x)$ defining C_2 is
$$\text{resultant}\big(y^2 - (u_1(x) + u_2(x))y^1 + u_1(x)u_2(x)y^0, \ (y - v_1(x)), \ y\big) =$$
$$\begin{vmatrix} 1 & -(u_1(x) + u_2(x)) & u_1(x)u_2(x) \\ 1 & -v_1(x) & 0 \\ 0 & 1 & -v_1(x) \end{vmatrix}.$$
This works out to precisely the product $(u_1(x) - v_1(x)) \cdot (u_2(x) - v_1(x))$, whose zero set is the set of x-values above which C_1 and C_2 intersect.

Remarkably, this extends to arbitrary finite products:
$$\text{resultant}\left(\prod_{i=1}^{m} (y - u_i(x)), \prod_{j=1}^{n} (y - v_j(x)), y\right) = \prod_{i,j=1}^{m,n} (u_i(x) - v_j(x)).$$

3.9. INTERSECTION MULTIPLICITY WITHOUT PARAMETRIZATIONS

For a proof, see [Walker], Theorem 10.10, p. 30. There is nothing in that proof requiring that u_i or v_j be polynomials, and in fact they could just as well be the fractional power series appearing in Theorem 3.1 on p. 54. Thus, in appropriate coordinates, let $p(x, y)$ and $q(x, y)$ be monic in y with degrees m and n. We may write

$$\text{resultant}(p(x, y), q(x, y), y) =$$

$$\text{resultant}\left(\prod_{i=1}^{m}(y - f_i(x^{\frac{1}{r_i}})), \prod_{j=1}^{n}(y - g_j(x^{\frac{1}{s_j}})), y\right),$$

so the resultant is the product of mn factors

$$\prod_{i,j=1}^{m,n}\left(f_i(x^{\frac{1}{r_i}}) - g_j(x^{\frac{1}{s_j}})\right). \tag{3.4}$$

We can use these observations to simplify computing the multiplicity of a point of intersection of two curves. Let C_1 and C_2 be curves defined by polynomials $p(x, y)$ and $q(x, y)$ of degree m and n. Assume pq has no repeated factors. (See Comment 3.4 on p. 63.) Choose p, q and coordinates so that

- $p(x, y)$ has the form $y^m + p_1(x)y^{m-1} + \cdots + p_m(x)$,
- $q(x, y)$ has the form $y^n + q_1(x)y^{n-1} + \cdots + q_n(x)$,
- $(0, 0) \in C_1 \cap C_2$,
- on the y-axis, $(0, 0)$ is the only point of $C_1 \cap C_2$.

Theorem 3.1 on p. 54 tells us that p and q have the following unique factorizations, with f_i and g_j analytic in a neighborhood of 0:

$$p(x, y) = y^m + p_1(x)y^{m-1} + \cdots + p_m(x) = \prod_{i=1}^{m}\left(y - f_i(x^{\frac{1}{r_i}})\right)$$

$$q(x, y) = y^n + q_1(x)y^{n-1} + \cdots + q_n(x) = \prod_{j=1}^{n}\left(y - g_j(x^{\frac{1}{s_j}})\right). \tag{3.5}$$

Mutually conjugate factors in each product geometrically define one and the same branch of C_1 or C_2, which may or may not go through the origin. Since $(0, 0) \in C_1 \cap C_2$, we know that the set of branches of C_1 through $(0, 0)$ is nonempty and the same is true of C_2. Let the factors of p

corresponding to branches through the origin be indexed by m', and the rest of p's factors indexed by m''. The factors indexed by m' have positive order at $x = 0$, while those indexed by m'' have order 0 at 0. Similarly, index the factors of q by either n' or by n'' according as their order at 0 is positive or 0.

Now we're ready to let (3.4) do its magic. The product

$$\text{resultant}(p(x,y), q(x,y), y) = \prod_{i,j=1}^{m,n} \left(f_i(x^{\frac{1}{r_i}}) - g_j(x^{\frac{1}{s_j}}) \right)$$

can be written in four parts:

$$\prod_{m',n'} \left(f_i(x^{\frac{1}{r_i}}) - g_j(x^{\frac{1}{s_j}}) \right) \times$$

$$\prod_{m',n''} \left(f_i(x^{\frac{1}{r_i}}) - g_j(x^{\frac{1}{s_j}}) \right) \times \prod_{m'',n'} \left(f_i(x^{\frac{1}{r_i}}) - g_j(x^{\frac{1}{s_j}}) \right) \times$$

$$\prod_{m'',n''} \left(f_i(x^{\frac{1}{r_i}}) - g_j(x^{\frac{1}{s_j}}) \right).$$

The first product in the second line has order 0 at $x = 0$ because each f_i has no constant term, while each g_j has a nonzero constant term, making $f_i - g_j$ have order 0. The same is true of the second product in that line. In the third line, f_i has a constant term, and so does g_j. They can't be the same, for if they were, then $f_i(0) = g_j(0) \neq 0$, and C_1 and C_2 would share a point on the y-axis other than the origin, contrary to the way we chose coordinates. Therefore the order of the product in the third line is 0. That means the order at 0 of the resultant (the whole big product) is the order of just the first line. Now each factor in the first line can be looked at as the result of substituting $y = f_i(x^{\frac{1}{r_i}})$ into $(y - g_j(x^{\frac{1}{s_j}}))$. This substitution can also be written as $q(x, f_i(x^{\frac{1}{r_i}}))$, where for each i there are r_i conjugate factors. By summing orders, we see from (3.5) that *the order of x in* $\text{resultant}(p(x,y), q(x,y), y)$ *equals the multiplicity of intersection of C_1 and C_2 at the origin.*

We can phrase the last italicized statement slightly more generally as a theorem:

Theorem 3.3. Let $p, q \in \mathbb{C}[x, y]$. Suppose pq has no repeated factors, and assume coordinates (x, y) have been selected so that above any x, there is at most one point of $C(p) \cap C(q)$. If $(x_0, y_0) \in C(p) \cap C(q)$, then the multiplicity of intersection of $C(p)$ and $C(q)$ at (x_0, y_0) is the order of resultant$(p(x, y), q(x, y), y)$ at x_0.

Example 3.15. Let's use the resultant to determine the intersection multiplicity at the origin of the two curves $C_1 = C(x^6 - x^2y^3 - y^5)$ and $C_2 = C(y^6 - y^2x^3 - x^5)$ shown in Figure 3.5 on p. 57. Entering

$$\text{resultant}(x^6 - x^2y^3 - y^5, y^6 - y^2x^3 - x^5, y)$$

into Maple gives

$$x^{27}(x^9 - 3x^6 - 9x^5 + 3x^3 - 18x^2 + 9x - 2).$$

Its order 27 agrees with what we found in Example 3.12.

Why, then, would we ever want to use parametrizations to compute intersection multiplicity? It's because they give the separate contributions of each pair of branches — any branch of C_1 through a point P with any branch of C_2 through P — and there are times this information can be useful. The resultant does not give that kind of detail, but it can greatly simplify getting the total intersection multiplicity at a point.

3.10 Bézout's theorem

Bézout's theorem is the multiplication theorem we promised at the beginning of this chapter, and is one of the cornerstones of our subject. Here's one formulation.

Theorem 3.4. (Bézout's theorem) Suppose that homogeneous polynomials $p(x, y, z)$ and $q(x, y, z)$ have degrees m and n, and that pq has no repeated factors. Then in $\mathbb{P}^2(\mathbb{C})$, $C_1 = C(p)$ and $C_2 = C(q)$ intersect in mn points, counted with multiplicity.

Our sketch of its proof will make it clear that Bézout's theorem can be alternately phrased this way:

Theorem 3.4a. (Bézout's theorem) Suppose that polynomials $p(x, y)$ and $q(x, y)$ have degrees m and n, and that pq has no repeated factors. Let C_1 and C_2 be the projective completions in $\mathbb{P}^2(\mathbb{C})$ of $C(p)$ and $C(q)$. Then C_1 and C_2 intersect in mn points, counted with multiplicity.

Comment 3.4. In each version of this theorem, the condition that pq have no repeated factors says two things:

- p cannot have any repeated factors; likewise for q. To illustrate the problem if they did, let $p = x^2$ and $q = y$. The zero set $C(x^2)$ is merely the y-axis, which intersects the x-axis $C(y)$ in 1 point, not $2 \cdot 1$.

- p and q cannot share any common factor. To see the problem if they did, let $p = x(x + y)$ and $q = x(x - y)$. In this case, $C(p) \cap C(q)$ contains the common component $C(x)$, the y-axis. Therefore the curves intersect in infinitely many points, not $2 \cdot 2$.

- Also, notice that Bézout's theorem implies that if C_1 and C_2 intersect in more than mn points, then the condition on pq cannot hold. If the degrees are m and n and neither polynomial has any repeated factors, then p and q must share a nonconstant factor, meaning that $C_1 \cap C_2$ includes a curve.

We'll use the following lemma, important in its own right, in sketching a proof of Bézout's theorem.

Lemma 3.1. If $C(p)$ and $C(q)$ in $\mathbb{P}^2(\mathbb{C})$ do not have a curve in common, then $C(p) \cap C(q)$ consists of finitely many points.

Proof. Suppose to the contrary that there are infinitely many points in $C(p) \cap C(q)$. Then we could select a projective line L in $\mathbb{P}^2(\mathbb{C})$ off which there are infinitely many points of $C(p) \cap C(q)$. Choose L as the line at infinity of $\mathbb{P}^2(\mathbb{C})$ and let (x, y) be coordinates in the corresponding affine plane. Let X be the finite set of values in the x-axis at which the polynomial resultant(p, q, y) vanishes. Let Y be the analogous set in the y-axis for resultant(p, q, x). The set $X \times Y$ is finite and $C(p) \cap C(q)$ is a subset of it.

In proving Bézout's theorem, we will also use

Theorem 3.5. Suppose $p(x, y, z)$ is homogeneous of degree m, and that $q(x, y, z)$ is homogeneous of degree n. Then

$$\text{resultant}\big(p(x, y, z), q(x, y, z), z\big)$$

is either 0 or homogeneous in x, y, z of degree mn.

Sketch of Proof. We will refer to Figure 1.7, so we reproduce it here:
Here, a_i and b_i are homogeneous in x, y of degree i, the resultant therefore being a polynomial $R(x, y)$. To show it's homogeneous of degree mn, replace x by tx and y by ty, giving $R(tx, ty)$. We want to show that this is

3.10. BEZOUT'S THEOREM

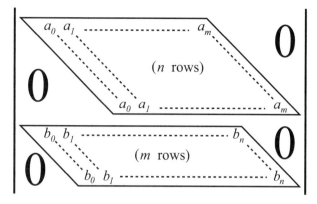

FIGURE 3.6.

$t^{mn} R(x, y)$. The essential idea of the argument is easy to follow when we run through a specific case such as $m = 5$ and $n = 3$. The general argument follows exactly the same pattern. In our case, the $(m + n) \times (m + n)$-matrix is 8×8, $p(x, y, z)$ is homogeneous of degree 5, and the powers introduced in each of the three a-rows by the replacement are respectively $t^0, \cdots, t^5$. Similarly, $q(x, y, z)$ is homogeneous of degree 3, and the powers introduced in each of the five b-rows by the replacement are respectively $t^0, \cdots, t^3$. So the replacement has multiplied the original matrix *elementwise* by the 8×8 array in Figure 3.7.

1	t	t^2	t^3	t^4	t^5		
	1	t	t^2	t^3	t^4	t^5	
		1	t	t^2	t^3	t^4	t^5
1	t	t^2	t^3				
	1	t	t^2	t^3			
		1	t	t^2	t^3		
			1	t	t^2	t^3	
				1	t	t^2	t^3

FIGURE 3.7.

Doing this in fact means the original resultant has been multiplied by $t^{mn} = t^{15}$, but this is not obvious from the figure! However, we can make the matrix much nicer by packing it with additional powers of t so that all entries in any column become uniform, as in Figure 3.8. Figure 3.8 is obtained from Figure 3.7 by multiplying the first $n = 3$ rows by $1, t$ and t^2, respectively — that is, by powers $t^0, \cdots, t^{n-1}$. The last $m = 5$ rows are multiplied by powers $t^0, \cdots, t^{m-1}$, respectively. We've therefore multiplied the

1	t	t^2	t^3	t^4	t^5		
	t	t^2	t^3	t^4	t^5	t^6	
		t^2	t^3	t^4	t^5	t^6	t^7
1	t	t^2	t^3				
	t	t^2	t^3	t^4			
		t^2	t^3	t^4	t^5		
			t^3	t^4	t^5	t^6	
				t^4	t^5	t^6	t^7

FIGURE 3.8.

matrix in Figure 3.7 by t-powers totaling $(1+2+3+4)+(1+2) = 13$:

$$\bigl(1+\cdots+(n-1)\bigr)+\bigl(1+\cdots+(m-1)\bigr) = \frac{(n-1)n}{2} + \frac{(m-1)m}{2}.$$

And the columns of the matrix in Figure 3.8? Their powers add up to 28 — that is to say, $\frac{(n+m-1)(n+m)}{2}$. Therefore the resultant $R(tx, ty)$ is $R(x, y)$ multiplied by t to the power $28 - 13 = 15 = 5 \cdot 3$, or generally,

$$\left[\frac{(n+m-1)(n+m)}{2}\right] - \left[\frac{(n-1)n}{2} + \frac{(m-1)m}{2}\right].$$

This last simplifies to mn The sketch of the proof is therefore complete.

Sketch of a Proof of Bézout's Theorem. We use the resultant. Since p and q share no nonconstant factors, C_1 and C_2 share no common curve. By Lemma 3.1, $C_1 \cap C_2$ consists of finitely many points. Therefore there is a line in $\mathbb{P}^2(\mathbb{C})$ missing all the intersection points. Choose coordinates (x, y, z) so that $y = 1$ defines this line and so that

$$\mathrm{resultant}\bigl(p(x, y, z), q(x, y, z), z\bigr),$$

which is a homogeneous polynomial in x and y of degree mn, has x^{mn} as a term. Now dehomogenize $\mathbb{P}^2(\mathbb{C})$, p, q and the resultant with respect to y. They are easily seen to have respective degrees m, n and mn. In the

3.10. Bézout's theorem

dehomogenized plane $\mathbb{C}^2$, further choose (x, y)-coordinates so that no two distinct intersection points lie on any line $x =$ constant. By our choice of coordinates, each zero of the resultant is the x-coordinate of exactly one point P of $C_1 \cap C_2$. The order of the zero is the intersection multiplicity at P, and the orders of the resultant at its zeros sum to its degree mn. Therefore the sum of all the multiplicities of intersection points of $C_1 \cap C_2$ is mn.

Let's look at some examples of Bézout's theorem.

Example 3.16. Consider again the two curves in Figure 3.5 on p. 57,

$$C_1 = C(x^6 - x^2y^3 - y^5), \ C_2 = C(y^6 - y^2x^3 - x^5).$$

Bézout's theorem tells us that C_1 and C_2 intersect in a total of $6 \times 6 = 36$ points. We can write resultant$(x^6 - x^2y^3 - y^5, y^6 - y^2x^3 - x^5, y)$ as

$$x^{27}(x - 2)(x^2 - x + 1)(x^6 + 3x^5 + 6x^4 + 8x^3 + 3x^2 - 3x + 1).$$

The origin in $\mathbb{C}^2$ accounts for 27 of the 36 points, and each of the other nine lies above a zero of

$$(x - 2)(x^2 - x + 1)(x^6 + 3x^5 + 6x^4 + 8x^3 + 3x^2 - 3x + 1).$$

One point is $(2, 2)$, which we'd see by extending Figure 3.5 rightward a little. Each of the other eight points has at least one non-real coordinate, so are not seen in $\mathbb{R}^2$.

Example 3.17. Let's revisit Example 2.11 on p. 42, the rational function $y = \frac{1}{(x^2-1)(x^2-4)}$. This can be written as the polynomial equation

$$y(x^2 - 1)(x^2 - 4) = 1,$$

with three views of its algebraic curve C appearing in Figure 2.12 on p. 43. Bézout's theorem tells us that in $\mathbb{P}^2(\mathbb{C})$, any projective line intersects the projective curve C in five points, counted with multiplicity. Figure 3.9 shows a series of seven lines $L_1, \cdots, L_7$. Let's test the theorem by following the history of the five intersection points.

For the line L_1, all five intersection points are finite, real, and distinct. As we rotate L_1 to L_2, we see in the bottom sketch that the leftmost intersection goes off to infinity and intersects the point of C there. This intersection is transverse (that is, the line isn't tangent to the curve there), and

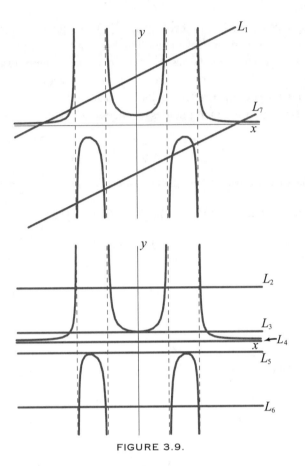

FIGURE 3.9.

the intersection multiplicity there is one. As L_2 moves downward toward L_3, the two middle intersection points approach each other, becoming a double point of intersection with L_3 when $y = \frac{1}{4}$. The line still intersects C transversally at P_∞, meaning the intersection multiplicity is one there, so L_2 intersects the curve in a total of five points. With further downward translation, the double point separates into two points with conjugate imaginary x-coordinates that go to infinity as the line approaches the x-axis. From Figure 3.9, we see that as L_3 translates downward toward the x-axis, the downward-moving line intersects C in two real points, and that these, too, race towards infinity as the line approaches the x-axis $= L_4$. Together with the single point at infinity, this comes to five points, one always remaining at infinity and four others approaching it. When the line reaches the x-axis itself, we see no points of intersection in the figure; all five are at

3.11. Bézout's Theorem Generalizes the FTA

the point at infinity of the x-axis, so C is tangent to the x-axis at its point at infinity. We can see the real part of this tangential intersection by homogenizing and dehomogenizing at $x = 1$ to get $z^5 = y(1 - z^2)(1 - 4z^2)$. Figure 3.10 shows the curve around P_∞. The curve intersects the original

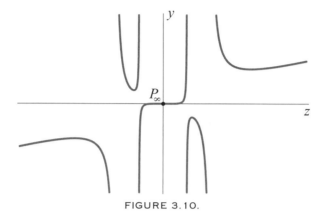

FIGURE 3.10.

x-axis (now named the z-axis) at the new origin, and we can compute the multiplicity by finding the resultant of the polynomials defining the view of C and what is now the z-axis:

$$\text{resultant}(z^5 - y(1 - z^2)(1 - 4z^2), y, y) = z^5.$$

This has order 5, so P_∞ indeed has multiplicity 5.

In Figure 3.9, as the moving horizontal line descends toward L_5, the point at infinity is now an intersection point of multiplicity 1. Two pairs of conjugate points (which we don't see) approach the two maximum points of the ∩-shapes appearing in the figure. In L_6 and L_7, everything is real and basically similar to L_2 and L_1.

3.11 Bézout's theorem Generalizes the Fundamental Theorem of Algebra

Geometrically, the Fundamental Theorem of Algebra can be looked at as a theorem counting the number of intersection points of two very special kinds of algebraic curves. One curve is the graph of a polynomial function of x and the other curve is the x-axis. Bézout's theorem is a far-reaching generalization of this. Here's a statement of the Fundamental Theorem of Algebra that suggests its kinship with Bézout's theorem:

Theorem 3.6. Let $p(x)$ be a nonzero polynomial. Then in $\mathbb{C}^2$, the curves $C_1 = C(y - p(x))$ and $C_2 = C(y)$ intersect in $\deg p \cdot \deg y$ ($= \deg p \cdot 1$) points, counted with multiplicity.

Notice that Theorem 3.6 is slightly more general than the algebraic statement guaranteeing a zero of any nonconstant polynomial, because if the polynomial p is a nonzero constant, its graph is a line not intersecting the x-axis in $\mathbb{C}^2$; p has degree 0, and its graph indeed intersects the x-axis in $\deg p \cdot \deg y = 0$ points in $\mathbb{C}^2$.

We can gradually relax the restrictive nature of the two polynomials defining the curves C_1 and C_2 in Theorem 3.6, moving from the Fundamental Theorem of Algebra to Bézout's theorem.

- The x-axis in Theorem 3.6 can be looked at as a "test line" that registers the guaranteed number $\deg p$ of intersection points with $C_1 = C(y - p(x))$. Even without Bézout's theorem, it's easy to see that any line $y = mx + b$ can equally well serve as a test line picking up $\deg p$ intersection points, as long as $p(x) \neq mx + b$. But what if m is ∞, corresponding to a "vertical line" in $\mathbb{C}^2$? Since in the Fundamental Theorem of Algebra $y = p(x)$ defines a function in $\mathbb{C}^2$, its graph satisfies the vertical line test, so any such line $x = $ constant intersects the graph in one point. Here Bézout's theorem comes to our aid, because it tells us that in $\mathbb{P}^2(\mathbb{C})$, the projective completions of any such line and of C_1 intersect in $\deg p$ points, counted with multiplicity. Figure 2.6 on p. 34 correctly suggests that this intersection point is at the end of the y-axis. Bézout's theorem tells us that the multiplicity of this intersection point is $\deg p - 1$.

- There is one other candidate for a test line in $\mathbb{P}^2(\mathbb{C})$ — the line at infinity. Of course the Fundamental Theorem of Algebra says nothing about this, but Bézout's theorem tells us this line can serve as a test line, too. Homogenize $y = p(x) = a_0 x^n + \cdots + a_n$ and dehomogenize at $y = 1$ to get $z^{n-1} = a_0 x^n + \cdots + a_{n-1} x z^{n-1} + a_n z^n$. Parametrizing the line at infinity by $\{x = t, z = 0\}$ and substituting into this dehomogenization gives $0 = t^n$, so in $\mathbb{P}^2(\mathbb{C})$, the curve $C_1 = C(y - p(x))$ intersects the line at infinity in one point of multiplicity $n = \deg p$.

- Bézout's theorem allows us to replace any of the above test lines by the graph of another polynomial $y = q(x)$ of degree, say, m. In $\mathbb{C}^2$, distinct graph-curves $C_1 = C(y - p(x))$ and $C_2 = C(y - q(x))$ can be massaged into the traditional picture. That is, subtracting

$q(x)$ from each of $p(x)$ and $q(x)$ produces a familiar-looking picture: $y = p(x) - q(x)$ intersecting the x-axis $y = 0$. Intersection points correspond in the two pictures, so C_1 and C_2 intersect in $\deg(p - q)$ points, counted with multiplicity. Usually this number isn't $\deg p \cdot \deg q$, but the two graphs intersect at infinity in one point of multiplicity $mn - \deg(p - q)$.

- To link with the full Bézout theorem, note that $y - p(x)$ and $y - q(x)$ are each two-variable polynomials of a special sort: the variables are separated and y occurs only with degree 1. Assuming that p and q are distinct (and monic) corresponds to assuming that $(y - p) \cdot (y - q)$ have no repeated factors. Notice that in massaging $C_1 = C(y - p(x))$ and $C_2 = C(y - q(x))$ as above into the traditional picture, we arrived at finding the zeros of $p(x) - q(x)$. This difference is just

$$\text{resultant}(y - p(x), y - q(x), y) = \begin{vmatrix} 1 & -p(x) \\ 1 & -q(x) \end{vmatrix} = p(x) - q(x).$$

Those special polynomials $y - p(x)$ and $y - q(x)$ are replaced by more general ones from $\mathbb{C}[x, y]$ in the full Bézout theorem.

3.12 An Application of Bézout's theorem: Pascal's theorem

Theorem 3.7. (**Pascal's theorem**) Let C be an ellipse in $\mathbb{P}^2(\mathbb{R})$, and let $P_1, \ldots, P_6$ be six numbered points on C. (The point after P_6 is P_1.) For each $i = 1, \ldots, 6$, successive points P_i, P_{i+1} define the side of a hexagon inscribed in C, as well as the projective line L_i through these two points. The projective line pairs extending opposite hexagon sides — that is, $\{L_1, L_4\}$, $\{L_2, L_5\}$ and $\{L_3, L_6\}$ — each define a point of intersection, as suggested by Figure 1.11 on p. 20. *These three points of intersection lie on a projective line.*

Proof of Pascal's theorem. The morphing sequence in Figure 1.12 on p. 21 suggests the idea of the proof. In this sequence, there are three αs for which the curve is reducible. When α is 0 or 1 the curve is the union of three lines and for some third α, the curve is an ellipse plus a line. It is this third α that delivers the proof's punch line, as we'll see. The three pairs of lines in Figure 1.11 generate the two triple-line curves: pick one line from each pair and let their union be the triple-line curve $C(p)$; let $C(q)$ be the union of the remaining three lines. Each of p and q is the product of three degree-1 polynomials, so both are cubics. Importantly, for every α, $(1-\alpha)p + \alpha q$ has degree three. (If p and q share a monomial of the same type $x^i y^{3-i}$ for some particular i, then since $(1-\alpha) + \alpha = 1$, each $(1-\alpha)p + \alpha q$ contains this monomial. If p and q share no monomial of the same type, they cannot cancel in $(1-\alpha)p + \alpha q$, so this still contains a degree three monomial.) Each line in $C(p)$ meets each line in $C(q)$ for a total of nine points of intersection. Figure 1.12 on p. 21 shows intermediate curves passing through all nine intersection points, as Property 1 on p. 19 guarantees. Property 2 says that for any point Q in the plane, there's some curve $(1-\alpha)p + \alpha q$ passing through Q. Here's the *coup de grâce*: Choose Q to be a point on the ellipse *other than* $P_1, \ldots, P_6$. The corresponding blend curve now contains *seven* points on the ellipse, while Bézout's theorem says that the ellipse (a quadratic) should intersect a blend curve (a cubic) in $2 \cdot 3 = 6$ points. Comment 3.4 on p. 63 tells us that the intersection of the ellipse and blend curve must include a curve. The ellipse is irreducible, so the blend curve must include the ellipse. Therefore the blend curve, which has degree three, is reducible and contains an ellipse (which has degree two), so the rest of the blend curve must be a line. Property 1 tells us the three intersection points lie on that line.

Comment 3.5.

- In proving Pascal's theorem, we chose a point Q on the ellipse other than $P_1, \ldots, P_6$, with Property 2 guaranteeing that for some α, the curve $(1-\alpha)p + \alpha q$ passes through Q. A little thought shows that the same α works for *every* point on the ellipse. That is, had we chosen a different Q on the ellipse, we'd still arrive at the same α that makes "the magic moment."

- Since C is an ellipse in $\mathbb{P}^2(\mathbb{R})$, in an affine view, C could be an ellipse, parabola or hyperbola.

- Because Pascal's theorem is projective, the theorem holds when two opposite hexagon sides are parallel. In this case the projective lines extending these sides meet at infinity. For a regular convex hexagon inscribed in a circle, all three intersection points lie on the line at infinity.

- Although the six points are numbered, they can be assigned any position on the ellipse. This means the hexagon may not be convex and its unextended sides may intersect.

3.12. An Application of Bézout's theorem: Pascal's theorem

- If two successive points are the same, then by a limiting process we can define the side as that point, and the line extending it, the projective line tangent to C at that point.

- The restriction to $\mathbb{P}^2(\mathbb{R})$ is arbitrary — all results generalize in a natural way to $\mathbb{P}^2(\mathbb{C})$.

- The proof may be rewritten slightly to cover degenerate conics such as two lines. Also, a direct proof not using Bézout's theorem is not difficult. Pappus of Alexandria (ca 290–350 AD) discovered a theorem of this sort, illustrated in Figure 3.11.

Theorem 3.8. (Pappus' hexagon theorem) If the ordered vertices of a hexagon alternately lie on two lines, then the three intersection points of the extended opposite-side pairs are collinear.

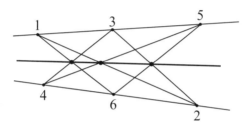

FIGURE 3.11.

CHAPTER 4

Topology of Algebraic Curves in $\mathbb{P}^2(\mathbb{C})$

4.1 Introduction

The gallery of real curves in Chapter 1 presented a wide range of behavior. It was so wide, we were led to ask "Where are the nice theorems?" We've already seen how broadening curves' living space to $\mathbb{P}^2(\mathbb{C})$ can lead to more unified results, Bézout's theorem in Chapter 3 being a prime example. But what about those real curves we met in Chapter 1 having more than one connected component? Or ones having mixed dimensions? Does working in $\mathbb{P}^2(\mathbb{C})$ perform its magic for cases like this?

Yes. In this chapter we'll see that individual curves in $\mathbb{P}^2(\mathbb{C})$ are generally much nicer and properties more predictable than their real counterparts. For example, we will show that every algebraic curve in $\mathbb{P}^2(\mathbb{C})$ is connected and that every irreducible curve is orientable. These are powerful theorems that help to smooth out the wrinkles in the real setting.

Most algebraic curves in $\mathbb{C}^2$ or $\mathbb{P}^2(\mathbb{C})$ are everywhere smooth. We will make this more precise in the next chapter, but any polynomial of a given degree with "randomly chosen" real or complex coefficients defines a real 2-manifold that is locally the graph of a smooth function. Such a curve in $\mathbb{P}^2(\mathbb{C})$ is therefore a closed manifold (a manifold having no boundary) that is orientable and thus has a topological genus. Remarkably, the genus depends only on the degree n of the defining polynomial:

$$g = \frac{(n-1)(n-2)}{2}.$$

A proof of this formula is sketched at the end of this chapter.

The first big result of this chapter is that algebraic curves are connected. We begin with some definitions.

4.2 CONNECTEDNESS

Definition 4.1. A topological space is *connected* if it isn't the disjoint union of two nonempty open sets.

The topological space S can be a subset of a larger space T. It is easy to check that the intersections of the open sets of T with S define a topology on S which we say is *inherited* from T. For us, any algebraic curve has a topology inherited from its ambient space, be it $\mathbb{R}^2$, $\mathbb{C}^2$, $\mathbb{P}^2(\mathbb{R})$, or $\mathbb{P}^2(\mathbb{C})$.

Example 4.1. Assign to the topological space $S = [-1, -\frac{1}{2}) \cup [0, 1]$ in $\mathbb{R}$ the topology inherited from the usual topology on $\mathbb{R}$. Then in S the subset $[-1, -\frac{1}{2})$ is open since it is $S \cap (-2, -\frac{1}{4})$. Likewise, $[0, 1]$ is open in S since $[0, 1] = S \cap (-\frac{1}{4}, 2)$. S is therefore the union of two nonempty open sets and thus is not connected. On the other hand, $S^* = [-1, 1]$ is connected, and breaking it into two pieces such as the disjoint union of the nonempty sets $[-1, 0) \cup [0, 1]$ means at least one of the pieces isn't open. Here $[-1, 0)$ is open in S^* but $[0, 1]$ is not.

Definition 4.2. A *path* in a topological space S is a continuous map from the unit interval $[0, 1]$ into S. The *endpoints* of any path are the images in S of 0 and 1 under the continuous map, and we say the endpoints are *connected by the path*.

Definition 4.3. A topological space S is *pathwise connected* if any two points P, Q of S can be connected by a path.

Theorem 4.1. *Any pathwise connected topological space is connected.*

For a proof, see [M-S], Theorem 7.30, p. 225. Pathwise connectedness is stronger than connectedness. For example, the closure in the real plane of the graph of $y = \sin \frac{1}{x}$ ($x > 0$) is connected but not pathwise connected. See [S-S], Section 4 for an informative discussion of various forms of connectedness.

4.3 Algebraic Curves are Connected

The major result in this section is

Theorem 4.2. Any algebraic curve in $\mathbb{P}^2(\mathbb{C})$ is connected.

Our proof will actually show more — that any algebraic curve $C(p)$ in $\mathbb{P}^2(\mathbb{C})$ is pathwise connected.

We may assume the nonconstant polynomial p is irreducible, because if Theorem 4.2 is true for p irreducible, then it's true for $p_1 p_2$, where p_1, p_2 are irreducible. This is because Bézout's theorem implies that $C(p_1)$ and $C(p_2)$ intersect in at least one point. Each curve $C(p_i)$ is pathwise connected, so there's a path from any point in $C(p_1)$ to the intersection point, and a path from that intersection point to any point in $C(p_2)$. The two paths together form a path from any point in $C(p_1)$ to any point in $C(p_2)$. An induction argument then shows that pathwise connectedness of irreducible curves implies pathwise connectedness of any curve.

Assuming now that p is irreducible, the proof comes down to showing that for any two points P, Q in $C(p)$, there is a path in $C(p)$ from P to Q. Choose a projective line in $\mathbb{P}^2(\mathbb{C})$ containing neither P nor Q and for convenience, continue to denote by p and $C(p)$ the polynomial and curve after dehomogenizing $\mathbb{P}^2(\mathbb{C})$ with respect to this line; let $\mathbb{C}^2$ be the corresponding dehomogenization of $\mathbb{P}^2(\mathbb{C})$. Choose coordinates in $\mathbb{C}^2$ so that p has the form

$$p(x, y) = y^n + a_1(x)y^{n-1} + \cdots + a_n(x).$$

Then over each point $x \in \mathbb{C}$ there lie n points of $C(p)$. Some may be multiple — in the language of intersection multiplicity, they're where lines $x = c$ intersect $C(p)$ with multiplicity > 1. They will play a crucial role in creating a path from P to Q.

Algebraically, how do we locate these multiple points? An example points the way: the unit circle in $\mathbb{R}^2$ defined by $p(x, y) = x^2 + y^2 - 1$. This curve intersects the line $x = 1$ in a double point; likewise for the line $x = -1$. These points are where the upper and lower semicircles meet, and are where points on vertical lines $x = c$ coalesce as $c \to 1$ or $c \to -1$. The left vertical tangent line is the limit of secant lines through two points coalescing to $(-1, 0)$; the right vertical tangent is the limit of secant lines through two points coalescing (1, 0). In algebraic terms, a tangent line is vertical when the partial derivative $p_y(x, y)$ is zero. Now $p_y(x, y) = 2y$

which is zero when $y = 0$ — that is, at the points $(1, 0)$ and $(-1, 0)$ of the circle. Similarly, for

$$p(x, y) = y^n + a_1(x)y^{n-1} + \cdots + a_n(x)$$

there are multiple points of intersection of $C(p)$ with the line $x = c$ whenever $p_y(c, y) = 0$. Since a multiple point is where $C(p)$ and $C(p_y)$ intersect, it is once again the resultant that supplies essential information, because resultant$(p(x, y), p_y(x, y), y)$ is zero at any point on the x-axis above which there are intersection points of multiplicity > 1. In addition to $p_y(P) = 0$, it may also happen that $p_x(P) = 0$. As we'll see in the next chapter, both derivatives vanishing at P is the criterion for P to be "singular," meaning *any* line through P intersects the curve in multiplicity > 1. The cross-point of the alpha curve is an example.

The polynomial

$$(-1)^{\binom{n}{2}}\text{resultant}\big(p(x, y), p_y(x, y), y\big)$$

is called the *discriminant of $p(x, y)$ with respect to y*. We denote it by $D_y(p)$ or sometimes by just D. When the discriminant polynomial is non-constant, its zero set in the x-axis of $\mathbb{C}^2$ consists of finitely many points, which we call *discriminant points*. They are also the zero set in the x-axis of

$$\text{resultant}\big(p(x, y), p_y(x, y), y\big).$$

This zero set leaves out one important point: the point at infinity of the x-axis. If we homogenize and dehomogenize so the original line at infinity becomes the "vertical axis through the new origin," it can happen that the new affine curve has a discriminant point at the new origin. An example is the parabola defined by $y^2 - x$, because homogenizing this to $y^2 - xz$ and dehomogenizing by setting $x = 1$ gives $y^2 - z$, and this curve in fact has a double point at the new origin. However, by rotating the (x, y, z)-coordinates a bit before dehomogenizing, we can eliminate this problem.

We assume from now on in this chapter that p is irreducible and that coordinates (x, y) have been chosen so that all discriminant points lie in the affine curve $C = C(p(x, y))$, with p still having the form

$$y^n + a_1(x)y^{n-1} + \cdots + a_n(x) \ . \tag{4.1}$$

Here's some intuition on how the discriminant arises in showing $C = C(p)$ is connected. In Figure 4.1, P and Q depict two arbitrary points of

4.3. ALGEBRAIC CURVES ARE CONNECTED

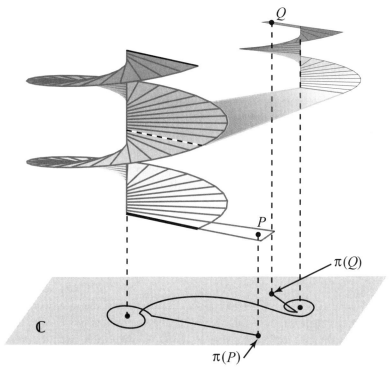

FIGURE 4.1.

C. The aim is to find a path in C connecting P and Q, and to do this we may need to climb "ramps." Discriminant points take center stage because the center of any ramp lies above some discriminant point. We climb a ramp in somewhat the same way that we climb a spiral staircase to go from one floor to the next in a building. Some readers may recognize this as winding around a branch point of a covering of the Riemann sphere to get from one sheet of the covering to another. In the circle example, there are two discriminant points $(\pm 1, 0)$, and in $\mathbb{C}^2$ the complex circle has ramps like those in Figure 4.1 at these two points. Each ramp allows us to go between floors, the upper semicircle being the real part of one floor, and the lower semicircle the real part of the other. In the general case of a polynomial $p(x, y)$ having the form in (4.1), there are n floors. Above any non-discriminant point in the x-axis $\mathbb{C}$ there are n distinct points, like n points directly above each other in successive floors of a building. The path sketched in Figure 4.1's complex x-axis depicts the shadow, or projection, of a journey in C from P to Q; π depicts the projection along the complex y-axis of $\mathbb{C}^2$ to the x-axis

$\mathbb{C}$. Starting on the floor containing P, we walk toward an available ramp, then climb it to another floor. Once there, we walk toward another ramp, climb it, then stroll over to Q. In general, we may need to use several ramps to create a path connecting P and Q.

The centers of the ramps in Figure 4.1 are where the tangent is "vertical," in the sense that the equation of the tangent line doesn't depend on y. In climbing either ramp in the circle example, after making two 360° turns, we're back on the original floor, the projected path winding twice about the discriminant point. However, for a curve like $y^3 = x$, vertical lines approaching the origin intersect the curve in three coalescing points, and a ramp would make three 360° turns about the origin before connecting to the original floor. For $y^n = x$, the ramp winds around n times. Topologically, the top and bottom edges of the ramp are identified, so the ramp is seamless. We could endlessly climb the ramp so that our shadow-point cycles round and round a closed curve winding n times around the ramp's discriminant point. In the figure, the left ramp shows both top and bottom edges, which are identified.

In general, we don't get to choose the location of the ramps or how many floors each staircase can access! That's determined by p. Although Figure 4.1 shows a path connecting P and Q, knowing only what we do at this stage, it's conceivable that there might not be enough staircases of the right type and location to allow us to path-connect two arbitrary points P and Q. For example in $\mathbb{C}^2$, a small neighborhood of the cross point of the alpha curve in $\mathbb{C}^2$ consists of two topological disks touching at a point. So in winding around either disk we remain on the same floor.

We carry out the proof by building upon the above topological ideas, employing some classical complex analysis together with the assumption that p is irreducible. Here are the concepts we'll use:

Definition 4.4. Let f be a function complex analytic in a domain $\mathfrak{D}$, a domain in $\mathbb{C}$ being any nonempty open connected set. Then the pair $(f, \mathfrak{D})$ is called an *analytic function element* or simply a *function element*.

Definition 4.5. Let $(f_1, \mathfrak{D}_1)$ and $(f_2, \mathfrak{D}_2)$ be function elements such that $\mathfrak{D}_1 \cap \mathfrak{D}_2 \neq \emptyset$ and such that f_1 and f_2 agree on $\mathfrak{D}_1 \cap \mathfrak{D}_2$. These two function elements are called *direct analytic continuations of each other*. For a sequence of function elements

$$(f_1, \mathfrak{D}_1), (f_2, \mathfrak{D}_2), \cdots, (f_m, \mathfrak{D}_m)$$

where successive function elements are direct analytic continuations of each other, any two function elements in this sequence are called *analytic continuations of each other*. The sequence of sets $(\mathfrak{D}_1, \mathfrak{D}_2, \cdots, \mathfrak{D}_m)$ with $\mathfrak{D}_i \cap \mathfrak{D}_{i+1} \neq \emptyset$ $(i = 1, \cdots, m-1)$ is called a *chain* of sets

In $\mathbb{C} \setminus \{\text{set of discriminant points}\}$, suppose that a chain of sets $\{\mathfrak{D}_i\}$ with $\{\mathfrak{D}_1\} = \{\mathfrak{D}_m\}$ winds around a discriminant point. The initial function element $(f_1, \mathfrak{D}_1)$ and the final function element $(f_m, \mathfrak{D}_1)$ may be different: f_1 may not agree with f_m on $\mathfrak{D}_1$. In the circle example, for instance, walking once in a small circle around either ramp leads to a sign change. In $\mathbb{R}^2$, that amounts to going from one semicircle to the other. On the other hand, walk in a tiny circle twice around either ramp, and then f_1 will agree with f_3 on $\mathfrak{D}_1$. Therefore, analytically continuing a function from one domain to another depends on the chain; the resulting functions are not in general uniquely determined.

We will use this far-reaching theorem:

Theorem 4.3. (Implicit Function Theorem) Let $P \in \mathbb{C}^2$. Suppose that $p(P) = 0$ and that $p_y(P) \neq 0$. Then within some sufficiently small neighborhood of P, the solutions of $p(x, y) = 0$ form the graph of a uniquely-defined function $y = f(x)$ analytic in that neighborhood.

For proofs of this basic result in higher dimensions, see for example [Fischer], Appendix 3, pp. 193–196, [Griffiths], Chapter I, § 9, or [Whitney], Appendix II § 3. This theorem is sometimes more fully referred to as the Holomorphic Implicit Function Theorem.

We will also use this definition:

Definition 4.6. Suppose a topological space S is pathwise connected. Let P and Q be any two points in S, and let γ_1 and γ_2 be any two paths in S each having P and Q as endpoints. If γ_1 can be continuously deformed into γ_2, with all intermediate paths remaining in S, then S is *simply connected*.

Now that we've presented some basic concepts, results and geometric motivation, let's get down to brass tacks and actually prove that $C(p)$ is connected! Again, $p(x, y)$ is irreducible of degree n with the form and coordinates assumed in (4.1) on p. 78. We begin by choosing any *non*discriminant point in $\mathbb{C}$ — let's call it a. Then
- There are n distinct points $\{a_1, \cdots, a_n\}$ of $C(p)$ above a.
- $p_y(a_i) \neq 0$, $(i = 1, \ldots, n)$.

Therefore, by the Implicit Function Theorem,

- For some sufficiently small open neighborhood U about a, the set $C(p) \cap \pi^{-1}(U)$ consists of the disjoint union of n graphs of functions $f_1 \ldots, f_n$ complex analytic on U.

We next use analytic continuation to significantly increase the size of U and of the disjoint analytic graphs lying above it. Here's the idea. First, $\mathbb{C}\setminus\{$set of discriminant points$\}$ isn't simply connected unless the set of discriminant points happens to be empty. But we can manufacture a region that *is* simply connected by removing more than just discriminant points, and then we can use analytic continuation to extend our n graphs, keeping them disjoint. The larger set we remove is a non-self-intersecting polygonal path ϕ that includes the discriminant points as vertices and whose final side is a ray. If necessary, add additional vertices so that the path misses $\pi(P)$ and $\pi(Q)$. Figure 4.2 shows an example. The heavy vertices denote discriminant points. The empty-circle vertex was added to make the path miss $\pi(P)$.

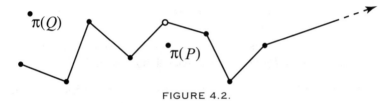

FIGURE 4.2.

Because $\mathbb{C} \setminus \phi$ is topologically an open disk with a radial line segment removed (and therefore still a topological disk), it is a simply connected region.

It follows from elementary complex analysis that since $\mathbb{C} \setminus \phi$ is simply connected, each function element $(f_1, U), \ldots, (f_n, U)$ extends to a function analytic on $\mathbb{C} \setminus \phi$. Let F_i denote the graph of f_i extended in this way. No graph F_i ever touches any other graph F_j because the n points above each point of $\mathbb{C} \setminus \phi$ are distinct. Thus the part of $C(p)$ over $\mathbb{C} \setminus \phi$ consists of n disjoint "analytic sheets" $F_1, \ldots, F_n$, each of which is topologically an open disk and therefore pathwise connected.

Now let $c_i \in \mathbb{C}$ be any one of the discriminant points. Choose a small circle in $\mathbb{C}$ about c_i, and a point (x, y) ($x \in \mathbb{C} \setminus \phi$, $y \in F_i$) above the circle. From this point, analytically continue $C(p)$ once around the circle, thus reaching some F_j (which may be F_i). This is reversible (just go around the circle in the other direction), so the association $F_i \to F_j$ is actually a permutation of the n objects $\{F_1, \ldots, F_n\}$. Choosing *any* discriminant point c_i creates a permutation of the set $\{F_1, \ldots, F_n\}$. All these permutations gen-

4.3. ALGEBRAIC CURVES ARE CONNECTED

erate a group Π, a subgroup of the symmetric group S_n of all permutations on n objects.

Our aim is to connect arbitrary points P and Q in $C(p)$. From what we've just said, and from the positions of $\pi(P)$ and $\pi(Q)$ in Figure 4.2, it's clear that P lies in some F_i and Q in some F_j. If $F_i = F_j$, we're done since the sheet, being a topological disk, is pathwise connected. So assume $F_i \neq F_j$. We've learned that by continuing analytically around little circles about discriminant points, we map from F_i to itself or to some other F_j. The essential question is this: is the group Π generated by all the above permutations *transitive* on $\{F_1, \ldots, F_n\}$? That is, if we specify arbitrary F_i and F_j, is there some combination of analytic continuations that will move us from floor F_i to floor F_j?

Under the action of Π, the sheets fall into orbits $\mathfrak{O}_1, \ldots, \mathfrak{O}_r$. Now Π acting transitively on $\{F_1, \ldots, F_n\}$ is the same as there being just one orbit — that is, $r = 1$. We prove that $r = 1$ by assuming $r > 1$ and deriving a contradiction. Therefore, let's say $\mathfrak{O}_1 = \{F_1, \ldots, F_s\}$, where $s < n$. The F_i, thought of as functions, can be added and multiplied pointwise in the usual way, so we can form elementary symmetric polynomials in them. Each symmetric polynomial remains unchanged under any permutation of the F_i, so each of them stays the same when we analytically continue it around discriminant points. This means each symmetric polynomial can be continued to all the discriminant points, thus becoming a meromorphic function on the Riemann sphere. By standard complex function theory, a function meromorphic on the Riemann sphere is a rational function. Furthermore, up to sign, these elementary symmetric polynomials are the coefficients of

$$(y - F_1)(y - F_2) \cdots (y - F_s) . \tag{4.2}$$

If we multiply all the rational-function coefficients by an lcm of their denominators and divide by a gcd of their numerators, the expression in (4.2) becomes a polynomial in both x and y (rather than merely analytic in x). Call this polynomial $q_1(x, y)$; its polynomial coefficients are relatively prime. Note that the equation $q_1(x, y) = 0$ is satisfied by each of the functions $y = F_i(x)$, $(i = 1, \ldots, s)$.

At this stage we've associated a $q_1 \in \mathbb{C}[x, y]$ with the orbit $\mathfrak{O}_1$. We can, in a similar way, associate a $q_j \in \mathbb{C}[x, y]$ with the orbit $\mathfrak{O}_j$, for each $j = 2, \ldots, r$. Now form the product

$$q(x, y) = q_1(x, y) q_2(x, y) \ldots q_r(x, y) .$$

Since we're assuming $r > 1$, this product has at least two factors, so $q(x, y)$ is reducible, has degree n, has relatively prime coefficients, and $q(x, y) = 0$ is satisfied by every $y = F_i(x)$, for $i = 1, \ldots, n$.

This last means that $q(x, y)$ is a constant multiple of $p(x, y)$, so q is irreducible. But because $r > 1$, we just said $q(x, y)$ is reducible! Thus $r > 1$ leads to a contradiction, so we must have $r = 1$. That is, the group Π acting on $\{F_1, \ldots, F_n\}$ generates a single orbit, meaning Π acts transitively on $\{F_1, \ldots, F_n\}$. We therefore can go from any floor to any other floor by analytically continuing around discriminant points. Since each floor is pathwise connected, $C(p)$ is pathwise connected, and therefore connected.

4.4 ORIENTABLE TWO-MANIFOLDS

Definition 4.7. A *real two-manifold*, or simply a *two-manifold*, is a connected Hausdorff space having a countable base of open neighborhoods, each topologically equivalent to an open disk of the Euclidean plane.

Definition 4.8. A topological disk in $\mathbb{R}^2$ can be *oriented*, in that about each of its points we can draw a small oriented topological loop in such a way that all the loop orientations are consistent. If a two-manifold has a covering by oriented open topological disks in such a way that orientations agree on all intersections, then the manifold is called *orientable*. If in an orientable two-manifold the assigned sense is always counterclockwise, then the manifold has been given a *positive orientation*. If the assigned sense is always clockwise, the manifold has been given a *negative orientation*.

Comment 4.1. An ordered basis (v_1, v_2) of $\mathbb{R}^2$ induces an orientation on $\mathbb{R}^2$ as follows: v_1 can be rotated into v_2 by either a counterclockwise or clockwise rotation through an angle less than π. If it's counterclockwise, we say the basis defines a positive orientation on $\mathbb{R}^2$; if clockwise, the basis defines a negative orientation. Typically in linear algebra, a geometric notion or statement has an algebraic counterpart, and vice-versa. In this case the basis defines a nonsingular matrix V whose ith column is v_i, and it can be shown that $\det V$ is positive when the basis induces a positive orientation and negative when the basis induces a negative orientation. For two ordered bases (v_1, v_2) and (w_1, w_2) with matrices V and W, there's a nonsingular A so that $W = AV$; (v_1, v_2) and (w_1, w_2) induce the same orientation if and only if $\det A$ is positive.

4.4. ORIENTABLE TWO-MANIFOLDS

Example 4.2. An open disk, sphere, torus and punctured plane are all examples of orientable two-manifolds. A Möbius strip with its boundary points removed, a Klein bottle, and the real projective plane $\mathbb{P}^2(\mathbb{R})$ are examples of non-orientable two-manifolds. Looking at $\mathbb{P}^2(\mathbb{R})$ as a disk with opposite points identified makes it easy to see why it isn't orientable: in Figure 4.3, start with a counterclockwise-oriented circle in the upper right part of the disk and gradually push it across the boundary. Since the disk's antipodal points are identified, the arc emerging on the other side, after it finally grows to a circle, has a clockwise orientation.

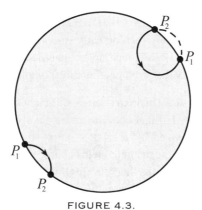

FIGURE 4.3.

Alongside the notion of orientability is *orientation-preserving map*, one which maps any oriented loop to one having the same orientation. This idea is important because a two-manifold is often constructed by gluing together open sets so that the transition map linking any pair of overlapping sets is continuous (or differentiable, or smooth, etc.). For example, consider the Möbius strip M constructed from the product $(0, 1) \times [0, 10]$ of two intervals by identifying the short sides after giving one of them a 180° twist. We can cover the region $(0, 1) \times (0, 10)$ with open unit squares, each given, say, a counterclockwise orientation. In using a final open unit square S to cover the short edge in the natural way, we see that $M \cap S$ minus the identified edge consists of two open sets having opposite orientations. The transition map between a pair of overlapping open sets must reverse orientation, so the Möbius strip is not orientable. In a two-manifold covered by open sets, if every transition map preserves orientation then the manifold can be given a consistent orientation and is therefore orientable.

4.5 Nonsingular Curves are Two-Manifolds

Intuitively, a curve like a line or a nondegenerate conic is smooth everywhere, while the part of the real alpha curve in some neighborhood of the crosspoint isn't. The part of the complex alpha curve in some $\mathbb{C}^2$-open neighborhood of the crosspoint looks like two open disks touching at just that one point. What about the origin of the cusp curve $y^2 = x^3$? Topologically, the part of the curve in a neighborhood of the origin is a disk, but in $\mathbb{C}^2$ it is contorted to the extent that in no rectangular coordinate system is a part around the origin the graph of an analytic function. Notwithstanding, at most points of any algebraic curve, the part of the curve about the point is actually more than merely topologically smooth. The Implicit Function Theorem tells us it is smooth in a complex analytic sense — at each of its points (x, y) it is locally the graph of a function complex-analytic at $x \in \mathbb{C}$.

The Implicit Function Theorem states a sufficient condition for a point of $C(p)$ to be smooth, but that condition is not a necessary one. For example, the parabola defined by $p = x - y^2$ is smooth everywhere, but $p_y(0, 0)$ isn't defined, so the theorem yields no information there. However, we could equally well state the theorem singling out x instead of y. In this form the theorem would tell us that if $p(0, 0) = 0$ and $p_x(0, 0) \neq 0$, then the part of the parabola around the origin is the graph of a function from a neighborhood of 0 in the complex y-axis to the complex x-axis. If at a point of $C(p)$ at least one p_x and p_y is nonzero, we could apply one or the other form of the theorem to conclude that at that point $C(p)$ is locally the graph of a complex-analytic function, and therefore smooth in that stronger sense. But we can't do this for the cusp curve $y^2 = x^3$ since for $p = y^2 - x^3$, both p_x and p_y are zero at the origin.

This leads us to the following basic definition.

Definition 4.9. Let p be irreducible. A curve $C(p) \subset \mathbb{C}^2$ is *nonsingular* at (x_0, y_0) if and only if $p(x_0, y_0) = 0$ and at least one of $p_x(x_0, y_0)$ and $p_y(x_0, y_0)$ is defined and nonzero. If $C(p)$ is nonsingular at each of its points in $\mathbb{C}^2$, then $C(p)$ is *nonsingular* as an affine curve. For a curve in the complex projective plane, if each of its points if nonsingular in some affine part, then we say the projective curve C is *nonsingular*.

We'll use Definition 4.9 in the next section.

4.6 Algebraic Curves are Orientable

The main fact we establish in this section is

Theorem 4.4. *Any irreducible algebraic curve $C \subset \mathbb{P}^2(\mathbb{C})$ is orientable.*

We give the main ideas of a proof; details appear in [Kendig 2]. The main task is establishing orientability for nonsingular curves; afterwards, it will be easy to extend the theorem to arbitrary irreducible curves. Therefore, assume for now that C is nonsingular. For any point $P \in C$, choose coordinates so that the part of C in some sufficiently small open neighborhood of P is the graph in $\mathbb{C} \times \mathbb{C}$ of a complex analytic function. In Figure 4.4, F_1 and F_2 depict two overlapping graphs, and for $i = 1, 2$, ϕ_i denotes

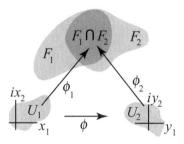

FIGURE 4.4.

an invertible analytic map from the open set U_i of the complex plane to F_i. ϕ is the composition $\phi_2^{-1} \circ \phi_1$, where its domain is restricted to the inverse image of $F_1 \cap F_2$ under ϕ_1. Regarding graphs F_1 and F_2 as typical overlapping neighborhoods in a covering $\{U_i\}$ of C, the question is, *does ϕ force orientations on U_1 and U_2 to agree?*

To decide, assume without loss of generality that the function maps origin to origin. Write $x = x_1 + ix_2$, $y = y_1 + iy_2$, as pictured in Figure 4.4. Then ϕ maps a point in U_1 into U_2 via $\phi(x_1, x_2) = (y_1, y_2)$. For $\epsilon > 0$ let $v_1 = (\epsilon, 0)$, $v_2 = (0, \epsilon)$ be an ordered basis of real (x_1, x_2)-space. Then $w_1 = \phi(v_1)$, $w_2 = \phi(v_2)$ is an ordered basis of real (y_1, y_2)-space. Since ϕ is invertible and analytic, for ϵ sufficiently small we can write

$$\begin{pmatrix} y_1 \\ y_2 \end{pmatrix} = J \cdot \begin{pmatrix} x_1 \\ x_2 \end{pmatrix} + \text{arbitrarily small higher order terms,}$$

where J is the nonsingular 2×2 Jacobian matrix $\left(\frac{\partial y_i}{\partial x_j}\right)$, $(i, j = 1, 2)$. From Comment 4.1 on p. 84, J preserves orientation if $\det(J)$ is positive. So *is* it? Let's compute:

$$\det(J) = \det\begin{pmatrix} \frac{\partial y_1}{\partial x_1} & \frac{\partial y_1}{\partial x_2} \\ \frac{\partial y_2}{\partial x_1} & \frac{\partial y_2}{\partial x_2} \end{pmatrix} = \frac{\partial y_1}{\partial x_1} \cdot \frac{\partial y_2}{\partial x_2} - \frac{\partial y_1}{\partial x_2} \cdot \frac{\partial y_2}{\partial x_1}.$$

Since ϕ is analytic, the Cauchy-Riemann equations $\frac{\partial y_1}{\partial x_1} = \frac{\partial y_2}{\partial x_2}$ and $\frac{\partial y_1}{\partial x_2} = -\frac{\partial y_2}{\partial x_1}$ hold. Substituting gives

$$\det(J) = \left(\frac{\partial y_1}{\partial x_1}\right)^2 + \left(\frac{\partial y_1}{\partial x_2}\right)^2.$$

This is indeed positive, so ϕ cannot reverse orientation. In a covering of a nonsingular algebraic curve C by positively-oriented open sets, the transition functions can never reverse orientation, so the orientation on C is consistent throughout. Therefore C is orientable.

Now suppose $C(p)$ has singular points, which means there are points on $C(p)$ at which both $\frac{\partial p(x,y)}{\partial x}$ and $\frac{\partial p(x,y)}{\partial y}$ are zero. Since p is assumed irreducible and each partial reduces the degree of p, the set of singular points of $C(p)$ is a proper subset of $C(p)$. Therefore by Lemma 3.1 on p. 64, the set of singular points is finite. If we remove these points from $C(p)$, what remains is a real two-manifold, so our proof for a nonsingular curve works for any curve with its singular points removed. At any singular point P, in a sufficiently small neighborhood of it, the part of any branch of $C(p)$ through the point is topologically an oriented open disk with the one point P missing. The consistent orientation within this punctured disk induces in a natural way an orientation at P consistent with the rest of the manifold, so the entire curve in $\mathbb{P}^2(\mathbb{C})$ is orientable.

4.7 The Genus Formula

In this section we sketch a proof of the following remarkable formula.

Theorem 4.5. Suppose an irreducible polynomial $p(x, y)$ of degree n defines a nonsingular curve $C \subset \mathbb{P}^2(\mathbb{C})$. The genus of C, as a closed orientable two-manifold, is

$$g = \frac{(n-1)(n-2)}{2}.$$

To prove this, we begin by selecting coordinates in an advantageous way. We use the same notation as in the proof of connectedness.

4.7. THE GENUS FORMULA

- Choose coordinates as on p. 78, meaning that p has the form

$$p(x, y) = y^n + a_1(x)y^{n-1} + \cdots + a_n(x),$$

and the point at infinity of the x-axis is not a discriminant point.

Above any discriminant point P there lies at least one point for which the intersection with the line $\pi^{-1}(P)$ has multiplicity two or greater. We want to insure that the multiplicity is in fact never more than two. Therefore:

- If necessary, apply a linear shear to linearly move the y-axis so that in the new coordinates, this holds: above each discriminant point P, among the n intersection points of the curve with the line $\pi^{-1}(P)$, exactly one has multiplicity ≥ 2 — all the others have multiplicity 1. In the real setting, a small shear amounts to a tilt, and the idea in this case is depicted in Figure 4.5. The two unbroken lines are parallel to the new y-axis, are

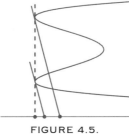

FIGURE 4.5.

tangent to the graph, and intersect the x-axis in distinct discriminant points.

- In shearing to move the y-axis but not the x-axis, Figure 4.5 suggests that there are infinitely many small such shears, each ensuring that all discriminant points are distinct. It turns out that for almost all of these changes in the y-axis, the intersection of $C(p)$ with a tangent line parallel to the new y-axis has multiplicity exactly two, rather than three or more. We assume that coordinates have been chosen so this holds. Therefore in these coordinates, just two points coalesce in determining the tangent line. The "almost all" phrase sheds light on Figure 4.1, p. 79. It says that the ramps there are typical, and that spiral staircases going up higher than those are the exception. As an example, consider the cubic curve defined by $y^3 - x$. The tangent line at $(0, 0)$ is the y-axis, and substituting its parametrization $\{x = 0, y = t\}$ into $y^3 - x$ gives t^3, which has order 3. Therefore that tangent line intersects the curve in multiplicity 3, and the staircase winds around one more time than either ramp Figure 4.1. But what about the intersection multiplicity with the tangent line at other points of the cubic?

To find out, let (a^3, a) $(a \neq 0)$ be any other point P of the cubic. For convenience, define new coordinates by $X = x - a^3$, $Y = y - a$ so that P is $(0,0)$ in the (X, Y) system. The slope at P of the cubic is $\frac{1}{3a^2}$, so the tangent line can be parametrized by $\{X = 3a^2 t, Y = t\}$. Now substitute this into $y^3 - x$ — that is, into $(Y + a)^3 - (X + a^3)$. After simplifying, this becomes $t^3 + 3at^2$, which has order 2 except when $a = 0$. Therefore except at one point, the tangent line of the cubic is defined by exactly two coalescing points. Correspondingly, the intersection multiplicity is 2. This gives the flavor of the general argument.

In the above coordinates (in which the point at infinity of the x-axis is not a discriminant point) we know that above each point of

$$\{\mathbb{C} \cup \infty\} \setminus \{\text{set of discriminant points}\}$$

there lie exactly n points of $C(p)$, and that above each discriminant point there lie exactly $n - 1$ points. Triangulate the topological sphere $\{\mathbb{C} \cup \infty\}$, and refine the triangulation until at most one discriminant point lies in the interior of any triangle. Add edges from a discriminant point to its triangle vertices so that the set of discriminant points becomes a subset of all triangulation vertices. Then π^{-1} lifts the triangulation to a triangulation of $C(p)$. Now all we need do is use Euler's formula $V - E + F = 2 - 2g$, where V, E and F denote the number of vertices, edges, and faces of a triangulated surface of genus g. Write Euler's formula as

$$g = 1 - \frac{V - E + F}{2}.$$

To apply it to our case, let V, E and F denote the number of vertices, edges, and faces of the above triangulated topological sphere $\{\mathbb{C} \cup \infty\}$. In the lifting of this to a triangulated curve $C(p)$, the number of faces is nF and the number of edges of nE. Over each discriminant point, exactly two points have coalesced to one, so the vertex count decreases by one over each discriminant point. Since the discriminant is the resultant of p (which has degree n) and a first derivative of p (which has degree $n - 1$), the degree of the resultant is $n(n - 1)$, meaning there are $n(n - 1)$ discriminant points. So the number of vertices is not nV, but rather $nV - n(n - 1)$. Substituting these numbers into the above displayed formula gives

$$g = 1 - \frac{n(V - E + F) - n(n - 1))}{2}.$$

4.7. The Genus Formula

For the sphere, $V - E + F$ is 2, so the genus of $C(p)$ is

$$g = 1 - \frac{n \cdot 2 - n(n-1)}{2} = \frac{2 - 2n + n(n-1)}{2} = \frac{n^2 - 3n + 2}{2}.$$

This last can be written in the more familiar form

$$g = \frac{(n-1)(n-2)}{2}.$$

CHAPTER 5

SINGULARITIES

5.1 INTRODUCTION

We have met curves that aren't everywhere smooth. For example in $\mathbb{R}^2$, the curve $y^2 = x^3$ has a cusp at the origin, and in a neighborhood of the origin the alpha curve $y^2 = x^2(x + 1)$ is ×-shaped. Each of these points is a *singularity* of the curve. The term "singular" connotes exceptional or rare. Within any particular complex affine or projective curve, singular points are indeed rare because there are only finitely many of them among the infinitely many points of the curve. A curve having no singularities is called *nonsingular*.

Singular points are rare in yet another way: most algebraic curves have no singularities at all! That is to say, if we randomly choose coefficients of $p(x, y)$, then $C(p)$ in $\mathbb{C}^2$ or $\mathbb{P}^2(\mathbb{C})$ is nonsingular. "Random" has the same meaning as in Chapter 1: a general polynomial $p(x, y)$ of degree n has finitely many coefficients, and since p and any nonzero multiple of it define the same curve, in randomly picking each of these finitely many coefficients, we may choose our dartboard to be the interval $(-1, 1) \subset \mathbb{R}$ for a polynomial in $\mathbb{R}[x, y]$, or from the unit disk about $0 \in \mathbb{C}$ for a polynomial in $\mathbb{C}[x, y]$.

In spite of their rarity, singular points can be found in curves defined by very simple polynomials, and understanding these special points can reveal quite a bit about the nature of algebraic curves in general. Important concepts in mathematics usually have both geometric and algebraic counterparts, and that's true of singular points.

5.2 Definitions and Examples

The notion of singular point of a curve has a number of equivalent definitions, some algebraic, others geometric. We now give a number of them for affine curves. In the next two sections, we extend our definitions to include projective curves.

Definition 4.9 on p. 86 says that an affine curve $C(p) \subset \mathbb{C}^2$ is nonsingular if and only if at each point (x_0, y_0), at least one of $p_x(x_0, y_0)$, $p_y(x_0, y_0)$ is nonzero. The condition ensures that the complex curve is everywhere smooth because we can apply the Implicit Function Theorem (Theorem 4.3 on p. 81), telling us that the curve is locally the graph of some analytic function. Definition 4.9 can be reworded as a characterization of singular point:

Definition 5.1. Let (x_0, y_0) be a point of the affine curve $C(p) \subset \mathbb{C}^2$. Assume p is nonconstant and has no repeated factors. The point (x_0, y_0) is *singular* if and only if $p_x(x_0, y_0) = p_y(x_0, y_0) = 0$.

Here is a more geometric characterization of singular point:

Definition 5.2. Suppose p is nonconstant and has no repeated factors, and let P be a point of the curve $C(p) \subset \mathbb{C}^2$. Let L denote a complex line through P and let the intersection multiplicity at P of L and $C(p)$ be m. The *order of P in $C(p)$* is the lowest value of m as L ranges over all lines through P. If $m > 1$, P is called a *singularity of order m*, a *singularity*, or a *singular point*.

Since one of the intersecting curves in the definition is a line, it is easy to recast the above definition algebraically. Let $P = (x_0, y_0) \in \mathbb{C}^2$. A typical complex line through P can be parametrized by

$$\{x = at + x_0, \ y = bt + y_0\}.$$

Definition 5.3. Let $P = (x_0, y_0)$ be a point of the curve $C(p) \subset \mathbb{C}^2$, where p is nonconstant with no repeated factors. If the minimum order m of $p(at + x_0, bt + y_0)$ over all pairs $(a, b) \in \mathbb{C}^2 \setminus \{(0, 0)\}$ is greater than 1, then P is called a *singularity of $C(P)$ of order m*, a *singularity*, or a *singular point*.

It is straightforward to check that these definitions don't depend on the choice of affine coordinates in $\mathbb{C}^2$.

Comment 5.1. Definition 5.3 is computationally useful, allowing us to find the order of a singularity in concrete cases. We illustrate this in examples below.

5.2. DEFINITIONS AND EXAMPLES

Another algebraic definition is based on the Taylor expansion of $p(x, y)$ about P.

Definition 5.4. Let $p(x, y)$ be nonconstant with no repeated factors. Expand p as a Taylor series about (x_0, y_0) as a sum of forms of increasing degree:

$$p(x, y) = \frac{1}{0!} p(x_0, y_0) + \frac{1}{1!} \left[p_x \cdot (x - x_0) + p_y \cdot (y - y_0) \right]$$
$$+ \frac{1}{2!} \left[p_{xx} \cdot (x - x_0)^2 + 2 p_{xy} \cdot (x - x_0)(y - y_0) + p_{yy} \cdot (y - y_0)^2 \right]$$
$$+ \frac{1}{3!} \left[p_{xxx} \cdot (x - x_0)^3 + \cdots \right] + \cdots,$$

where each partial is evaluated at (x_0, y_0). The point (x_0, y_0) is on the curve $C(p) \subset \mathbb{C}^2$ if and only if $p(x_0, y_0) = 0$, and is nonsingular there if at least one first partial is nonzero there. If for $m > 1$, some $(m + 1)^{\text{st}}$ partial is nonzero at (x_0, y_0) but all lower-order partials there are 0, then (x_0, y_0) is a singularity of $C(p)$ of order m.

Comment 5.2. In Definition 5.4, the initial form consists of the lowest-degree terms in the expansion, and this initial form defines the tangent cone to $C(p)$ at (x_0, y_0).

Example 5.1. The polynomial $p = y^2 - x^2(x + 1)$ defines an alpha curve with its cross point at the origin. Let the parametrization of a complex line through the origin be $\{x = at, y = bt\}$ and substitute it into p, getting $p(at, bt) = (b^2 - a^2)t^2 - a^3 t^3$. The lowest order of $p(at, bt)$ is 2 (occurring when $a^2 \neq b^2$), so the order of the origin P in the curve is 2 and therefore P is a singular point. The order is 3 when $b = \pm a$, which occurs when the line is tangent to the curve.

Definition 5.5. A singularity P of order m is called *ordinary* if exactly m distinct (complex) lines intersect $C(p)$ at P with multiplicity greater than m. In algebraic terms, there are exactly m parametrizations of lines that, when substituted into $p(x, y)$, result in an order greater than m.

Intuitively, a singularity of order m is ordinary if there are m distinct lines tangent to $C(p)$ at P. That is, the tangent cone at P consists of m distinct lines, with no double or other multiple lines.

Definition 5.6. A *node* is an ordinary singularity of order 2.

A node can be considered the simplest kind of singularity. The alpha curve's singularity is a node, while a cusp singularity is not even ordinary.

Example 5.2. $p = y^2 - x^3$ defines a curve with a cusp at the origin. Substituting into p the parametrization $\{x = at, \ y = bt\}$ gives $b^2t^2 - a^3t^3$ which has minimum order 2, occurring whenever $b \neq 0$. This singularity is not ordinary since there are not exactly 2 distinct lines intersecting $C(p)$ at the origin with multiplicity greater than 2; there is only one such line, the one corresponding to $b = 0$ yielding the parametrization $\{x = at, \ y = 0\}$ which defines the x-axis.

Example 5.3. We've seen from (1.2) on p. 11 that $(x^2+y^2)^{n+1} = [\Re(z^n)]^2$ defines a $2n$-petal rose when n is even. When $n = 2$, this equation becomes $(x^2 + y^2)^3 = (x^2 - y^2)^2$ and defines a four-leaf rose whose real portion is depicted in Figure 5.1. In the real plane we see self-intersections at the origin. It's easy to check that $p(at, bt)$ has order 4 except when $b = \pm a$, and for these exceptions the order becomes 6. The real figure correctly suggests that complex tangent lines through the origin correspond to $b = \pm a$ — that is, lines of slope ± 1. The lowest-degree form of the defining polynomial is $(x^2 - y^2)^2 = (x + y)^2 \cdot (x - y)^2$; the tangent lines defined by $x \pm y = 0$ are each double. The four lines intersecting the singularity in multiplicity greater than 4 aren't distinct, so the singularity is not ordinary.

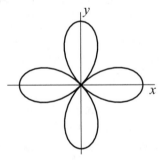

FIGURE 5.1.

Example 5.4. We can create singularities of any order N at any point. For example, $p = \prod_{k=1}^{N} ((y - y_0) - k(x - x_0))$ defines the union of N distinct lines through (x_0, y_0). The order of

$$p(at + x_0, bt + y_0) = \prod_{k=1}^{N} (bt - k(at))$$

is N except when the line $\{x = at + x_0, \ y = bt + y_0\}$ is one of the original N lines. In that case $p(at + x_0, bt + y_0)$ becomes the zero function which by definition has order ∞. Therefore (x_0, y_0) is an ordinary singular point of $C(p)$ of order N.

For any polynomial $p(x, y)$ without repeated factors, a necessary and sufficient condition for $C(p)$ to be nonsingular at the point $P = (0,0)$ is that $p(x, y)$ contain at least one first-order term, since otherwise $p(at, bt)$ has order at least 2. This can be restated using partial derivatives: at least one of $p_x(P)$, $p_y(P)$ is nonzero, which is what Definition 5.1 expresses.

5.3 Singularities at Infinity

Definition 5.3 of singular point applies to a point P of an affine curve in $\mathbb{C}^2$. What if P is on the line at infinity? One solution is to choose new coordinates so that after homogenizing and appropriately dehomogenizing, the new line at infinity misses P, and then any of Definitions 5.1–5.4 can be used to decide whether a point is singular, and any of Definitions 5.2–5.4 can be used to find its order. We leave it as an exercise to show that this approach is independent of the above choice of coordinates

Example 5.5. Consider the cubic $y = x^3$. Its branches approach the point P at the "end" of the y-axis. Is this cubic singular or nonsingular there? Homogenizing $y = x^3$ and then dehomogenizing at $y = 1$ yields the cusp curve $z^2 = y^3$, and P is a singularity of order 2.

Example 5.6. An argument like that in the above example shows that any curve defined by $y^2 = (x - r_1)(x - r_2)(x - r_3)$ is nonsingular at infinity.

5.4 Nonsingular Projective Curves

Sometimes it's important to know when a projective algebraic curve is *nonsingular* in the sense that it has no singularities, even at infinity. There are two main approaches to deciding. One is piecemeal, using affine views; the other is a "global" approach, which makes the determination by directly using a homogeneous polynomial.

A Piecemeal Strategy. Homogenize the polynomial defining the curve, then dehomogenize at each of $x = 1$, $y = 1$, $z = 1$. It is easy to check that these three affine views cover every point of $\mathbb{P}^2(\mathbb{C})$. If, by using any of Definitions 5.1 - 5.4, we can show there are no singular points in any of the three views, then the projective curve is nonsingular. Here's an example:

Example 5.7. For any positive integer n, the projective Fermat curve $C \subset \mathbb{P}^2(\mathbb{C})$ defined by $x^n + y^n = 1$ is nonsingular. This is obvious when $n = 1$, for then the curve is a line. Therefore assume $n \geq 2$ and homogenize the polynomial $p = x^n + y^n - 1$ to $p = x^n + y^n - z^n$. Dehomogenizing at $x = 1$ yields $1 + y^n - z^n$, and the partials $p_y = ny^{n-1}$ and $p_z = nz^{n-1}$ are simultaneously zero only at the origin of the (y, z)-plane. However, the origin doesn't lie on C, so C is nonsingular at each of its points in this plane. Dehomogenizing at $y = 1$ and using a similar argument shows that the affine curve in the (x, z)-plane is nonsingular at each of its points. Dehomogenizing at $z = 1$ likewise tells us there are no singular points in the (x, y)-plane. Therefore the projective Fermat curve $C \subset \mathbb{P}^2(\mathbb{C})$ is nonsingular. We will redo this example in Example 5.8, where the "global" approach is more efficient.

A "Global" Approach. Here, $h(x, y, z)$ is a homogeneous polynomial defining the projective curve $C(h) \subset \mathbb{P}^2(\mathbb{C})$. This polynomial could, for example, be given the homogenization of a polynomial $p(x, y)$ defining an affine curve $C(p) \subset \mathbb{C}^2$.

Comment 5.3. Suppose h is a nonconstant homogeneous polynomial with no repeated factors. If $C(h) \subset \mathbb{P}^2(\mathbb{C})$ is nonsingular, then h must be irreducible. This can be shown by contradiction: suppose h is reducible, say $h = h_1 h_2$, h_i homogeneous. Then by Bézout's theorem, there exists a point $P \neq (0, 0, 0)$ in $C(h_1) \cap C(h_2)$, so $h_1(P) = h_2(P) = 0$. Then $h_x(P) = h_{1x}(P)h_2(P) + h_1(P)h_{2x}(P) = 0 + 0$. Similarly, $h_y(P) = h_z(P) = 0$, so P is a singular point in $C(h)$.

Definition 5.7. Let $h(x, y, z)$ be nonconstant and homogeneous with no repeated factors, and let (x_0, y_0, z_0) be a nonorigin point of $\mathbb{C}^3$. Expand h as a Taylor series about (x_0, y_0, z_0) as a sum of forms of increasing degree:

$$h(x, y, z) = \frac{1}{0!} h(x_0, y_0, z_0)$$

$$+ \frac{1}{1!} \left[h_x \cdot (x - x_0) + h_y \cdot (y - y_0) + h_z (z - z_0) \right]$$

$$+ \frac{1}{2!} \left[h_{xx} \cdot (x - x_0)^2 + 2h_{xy} \cdot (x - x_0)(y - y_0) + \cdots + h_{zz} \cdot (z - z_0)^2 \right]$$

$$+ \frac{1}{3!} \left[h_{xxx} \cdot (x - x_0)^3 + \cdots \right] + \cdots,$$

where each partial is evaluated at (x_0, y_0, z_0). A nonorigin point (x_0, y_0, z_0) is on the curve $C(h) \subset \mathbb{P}^2(\mathbb{C})$ if and only if $h(x_0, y_0, z_0) = 0$, and is nonsingular there if some first partial is nonzero there. If for $m > 1$, some $(m + 1)^{\text{st}}$ partial is nonzero at (x_0, y_0, z_0) but all lower-order partials there are 0, then (x_0, y_0, z_0) *is a singularity of* $C(h)$ *of order* m.

5.5. SINGULARITIES AND POLYNOMIAL DEGREE

Comment 5.4. In Definition 5.7, the initial form consisting of the lowest-degree terms in the expansion defines the tangent cone to $C(p)$ at (x_0, y_0, z_0).

Definition 5.7 leads to this definition of nonsingular projective curve:

Definition 5.8. Let $h(x, y, z)$ be nonconstant and homogeneous. If at each nonorigin point of $C(h) \subset \mathbb{P}^2(\mathbb{C})$ not all first partials of h vanish, then $C(h)$ is nonsingular.

Example 5.8. Look at the Fermat curve again. Its defining homogeneous polynomial is $h = x^n + y^n - z^n$, and the three first partials are

$$h_x = nx^{n-1}, \quad h_y = ny^{n-1}, \quad h_z = -nz^{n-1}.$$

For $n \geq 1$, these are all nonzero at each non-origin point of $C(h)$, so the projective Fermat curve is nonsingular.

Example 5.9. For any $n \geq 0$, $y - x^n$ defines a projective curve $C \subset \mathbb{P}^2(\mathbb{C})$. For which values of n is C nonsingular? If n is either 0 or 1, C is a projective line, which is nonsingular. We can use Definition 5.8 to show that the only other value of n making C nonsingular is $n = 2$. Let's therefore assume $n \geq 2$, and form the homogenization $h(x, y, z) = yz^{n-1} - x^n$. Its first partials are

$$h_x = -nx^{n-1}, \quad h_y = z^{n-1}, \quad h_z = (n-1)yz^{n-2}.$$

Supposing all three partials are zero at a point (x_0, y_0, z_0), what can be said about that point? $h_x = -nx^{n-1}$ tells us that x_0 must be 0. $h_y = z^{n-1}$ tells us that z_0 must be 0. $h_z = (n-1)yz^{n-2}$ tells us that if $n = 2$, then $h_z = y$, so y_0 must be 0. So for $n = 2$, all three partials vanish only at $(0, 0, 0)$, which doesn't define a point of $\mathbb{P}^2(\mathbb{C})$. Therefore the complex projective parabola is nonsingular. If on the other hand $n > 2$, then $h_z = (n-1)yz^{n-2}$ can be zero when $y = 1$, and in that case the complex 1-space through $(0, 1, 0)$ represents a singular point of C.

5.5 SINGULARITIES AND POLYNOMIAL DEGREE

We begin with a little intuition. For curves in $\mathbb{R}^2$, singular points are often "self-intersections" where in tracing the curve we revisit a point already plotted. For example, if an ant walks in a smooth path along an alpha curve, it heads in a different direction when next passing through the self-intersection. With algebraic curves, such turning around imposes requirements on the defining polynomial's degree. This is familiar when the curve is the graph of a polynomial $y = p(x)$: intersections with the complex x-axis count the number of times the graph has "reversed course," which is one way of looking at the Fundamental Theorem of Algebra. (In fact, as we saw in Chapter 3, any line $y = mx$ can be used to keep score.) Intuitively, to plot a curve $C(p)$ with many singular points, we may need to turn around or reverse course frequently, so p cannot have a low degree. Put differently, for a fixed degree, the possible number of singular points is limited.

Here are some examples. If the degree of $p(x, y)$ is 1, then $C(p)$ is a line and there are no singularities. If the degree is 2 and p has no repeated factors, then the only singularity possible is in a degenerate conic consisting of two crossing complex lines; ellipses, parabolas and hyperbolas are all nonsingular. If C of degree n is the union of n randomly-chosen lines in $\mathbb{R}^2$ or $\mathbb{C}^2$, then C has $n(n-1)/2$ singular points since from n lines there are $n(n-1)/2$ ways of selecting unordered line-pairs, and any pair intersects in a singular point. It turns out that singular points of higher order place a greater demand on the degree of p. In Example 5.4 on p. 96 it took n lines and therefore an equation of degree n to create an ordinary singularity of order n. Therefore n lines can form $n(n-1)/2$ nodes but only one one ordinary singularity of order n. These two examples are related: by appropriately translating each of the n randomly-selected lines, we can make the $n(n-1)/2$ nodes draw ever closer to each other, in the limit all coalescing into one ordinary singularity of order n. In this way **we may think of an ordinary singularity of order n as being composed of $n(n-1)/2$ nodes.**

Comment 5.5. A loose analogy can be made with physical atoms. Split or fragment a heavy atom (think of one high-order singularity), and we'll get several lighter particles (think of several nodes), plus extra energy that held the heavy atom together.

The above curves made from lines are not irreducible, but it turns out they can be made so by adding sufficiently high even powers of x and y. (See the section "Designer Curves" starting on p. 22.) This exemplifies a general fact. Looking at singularities of only irreducible curves imposes a yet greater degree cost: in making the curve irreducible by adding high even powers of x and y, we increase the degree. On the other hand, suppose a polynomial p with no repeated factors can be written as a product qr of smaller-degree factors. By Bézout's theorem, $C(q)$ and $C(r)$ in $\mathbb{P}^2(\mathbb{C})$

5.5. SINGULARITIES AND POLYNOMIAL DEGREE

intersect in $\deg(q) \cdot \deg(r)$ points. As the argument in Comment 5.3 on p. 98 shows, any intersection point P is singular because both p_x and p_y are zero there. The following two theorems sum up much of what we've just said. (For proofs, see [Fischer], p. 49.)

Theorem 5.1. An *arbitrary* algebraic curve of degree n in $\mathbb{P}^2(\mathbb{C})$ can have at most $n(n-1)/2$ singularities.

Theorem 5.2. An *irreducible* algebraic curve of degree n in $\mathbb{P}^2(\mathbb{C})$ can have at most $(n-1)(n-2)/2$ singularities.

In each case, there exist curves having the maximum possible number of singularities. For arbitrary curves, the union of n randomly-chosen lines does it. For irreducible curves, the curve having parametrization

$$\{ x = \cos\left((n-1)\arccos t\right), \, y = \cos\left(n \arccos t\right) \} \quad (t \in \mathbb{C})$$

turns out to be algebraic of degree n, irreducible, and its singularities consist of exactly $(n-1)(n-2)/2$ nodes, all of which are real. For a proof and discussion, see [Fischer], 3.9, pp. 50–57. When restricted to the square $[-1, 1] \times [-1, 1]$, the real portions of these curves look like the Lissajous figures mentioned in Chapter 1, but without this restriction they can be unbounded in the real plane. For $n = 8$, for example, the parametrization defines a curve of degree 8 with $(8-1)(8-2)/2 = 21$ nodes, illustrated at the left in Figure 5.2.

For small n, it is quite feasible to locate all $(n-1)(n-2)/2$ nodes guaranteed by Theorem 5.2. For example, the real Lissajous figure shown

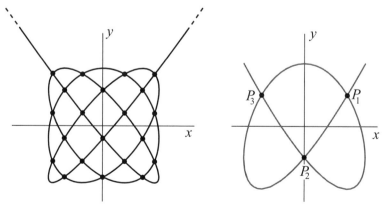

FIGURE 5.2.

at the right in Figure 5.2 is parametrized by

$$\{x = \cos(n - 1)t, \ y = \cos nt\},$$

with $t \in \mathbb{R}$ and $n = 4$. If we set $T = \cos t$, then multiple-angle formulas give

$$x = \cos(n-1)t = \cos 3t = 4T^3 - 3T = a(T),$$
$$y = \cos nt = \cos 4t = 8T^4 - 8T^2 + 1 = b(T).$$

Now $\cos n(n-1)t$ can be viewed in two ways: as the cosine of n times the argument $(n-1)t$, or as the cosine of $n-1$ times the argument nt. These are the same, so in terms of the polynomials a and b, we can write $a(b(T)) = b(a(T))$ — or, what is the same, $a(y) = b(x)$. For $n = 4$, this becomes

$$4y^3 - 3y = 8x^4 - 8x^2 + 1.$$

Self-intersections occur at points

$$(x, y) = \left(\cos \frac{k\pi}{n}, \ \cos \frac{k\pi}{n-1}\right)$$

whenever k is relatively prime to both n and $n-1$. When $n = 4$, this happens for $k = 1, 2, 5$ before we begin to cycle through self-intersections. These values of k give locations of the three nodes shown in right sketch:

$$P_1 = \left(\frac{\sqrt{2}}{2}, \frac{1}{2}\right), \quad P_2 = \left(0, \frac{-1}{2}\right), \quad P_3 = \left(\frac{-\sqrt{2}}{2}, \frac{1}{2}\right).$$

5.6 Singularities and Genus

As we've noted, the degree of an irreducible polynomial $p(x, y)$ affects how much $C(p)$ twists and turns in $\mathbb{P}^2(\mathbb{C})$, with the presence of singularities on an irreducible curve requiring a certain minimum amount of twisting and turning. Theorems 5.1 and 5.2 make this quantitative. For example, if the curve $C(p)$ has 100 singularities, then $p(x, y)$ must have degree at least 15 — no curve of smaller degree can have that many singularities. That much twisting and turning is required to get such a large number of singularities. The greater the number of singularities, the greater the minimum degree, and irreducibility further increases the degree requirement: if $C(p)$ is irreducible and has a million singularities, then the degree of p must be at least 1,416.

5.6. Singularities and Genus

Singularities are not the only thing requiring twisting and turning and therefore a sufficiently large degree. So does genus. A smooth curve with many holes can't possibly have all the implied contortions without a sufficiently high degree. The basic genus formula makes this quantitative: a nonsingular curve of degree n has genus $(n-1)(n-2)/2$. So both the genus and the number of singularities make demands on the degree. Now there certainly are curves having both high genus and many singularities; do these two actually compete for degree? To answer this we need the following theorem that connects topology with singularities.

Theorem 5.3. Topologically, any irreducible curve $C(p) \subset \mathbb{P}^2(\mathbb{C})$ with singularities is a compact oriented 2-manifold upon which this operation has been performed: identify finitely many of its points to finitely many points.

Example 5.10. As an example of identifying finitely many points to finitely many points, select three points on a rubber torus and pull them together to a single point. Select another five points and similarly identify them to another single point.

We extend the notion of genus of a compact oriented 2-manifold in the following way:

Definition 5.9. Suppose a compact oriented 2-manifold M has genus g. We say that identifying finitely many points of M to finitely many points leaves the genus unchanged. This new topological object is said to have *genus g*.

Here's the big question: suppose we start with a nonsingular curve $C(p) \subset \mathbb{P}^2(\mathbb{C})$ of degree n and genus g and continuously modify p so that at some stage $C(p)$ gains a singularity. Can we "trick" the curve so it gains the singularity without decreasing the genus? Or is it more a zero-sum game, in that gaining a singularity necessitates giving up a hole or two? Let's explore this idea with some examples.

Example 5.11. Consider the curve defined by $y^2 = (x-r_1)(x-r_2)(x-r_3)$. Let's assume that all three r_i are distinct, say $y^2 = (x+1)x(x-1)$ — that is, $y^2 = x^3 - x$. Figure 5.3 depicts its real portion. It is straightforward to check that this degree-3 curve in $\mathbb{P}^2(\mathbb{C})$ is nonsingular. Therefore it is a closed oriented 2-manifold having genus $g = (3-1)(3-2)/2 = 1$, making the curve a topological torus. The real portion appearing in Figure 5.3 reveals only a tiny part of the entire complex curve. To see it all, we'd need to look in real 4-space. Fortunately, the part in one particular real

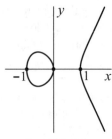

FIGURE 5.3.

3-dimensional slice is especially suggestive. It's a real space curve, and we get equations for it this way: Write $x = x_1 + ix_2$ and $y = y_1 + iy_2$, then substitute these into $y^2 = (x+1)x(x-1) = x^3 - x$, getting

$$(y_1 + iy_2)^2 = (x_1 + ix_2)^3 - (x_1 + ix_2).$$

Equating real and imaginary parts splits this into

$$y_1^2 - y_2^2 = x_1^3 - 3x_1x_2^2 - x_1, \quad 2y_1y_2 = 3x_1^2x_2 - x_2^3 - x_2 \qquad (5.1)$$

Each of these two equations in four real variables defines a 3-surface in $\mathbb{R}^4$, and together they determine a 2-surface $C(p) \subset \mathbb{C}^2 = \mathbb{R}^4$. We now additionally restrict to the 3-space $x_2 = 0$, which further cuts down the 2-surface to a real curve, and this is easily visualized in (x_1, y_1, y_2)-space. Now $x_2 = 0$ reduces the imaginary-part equation in (5.1) to $y_1 y_2 = 0$, so either $y_1 = 0$ or $y_2 = 0$. When $y_2 = 0$, the equation becomes the original equation restricted to the real plane, so once again we get the real curve in Figure 5.3. When $y_1 = 0$, the resulting real-part equation becomes $-y_2^2 = x_1^3 - x_1$, which is essentially the reflection of Figure 5.3's curve about the y_1 axis, but drawn in the (x_1, y_2)-plane. The two appear together in Figure 5.4 (a).

A nonzero value for x_2 defines a parallel copy of the 3-space $x_2 = 0$ and again leads to the intersection of three 3-surfaces in $\mathbb{R}^4$. Each codimension-1 surface cuts down the dimension by one and we end up with a real curve for each value of x_2. All these curves fit together, their closure in $\mathbb{P}^2(\mathbb{C})$ forming the topological torus, depicted in Figure 5.4 (b). Four rings are sketched on the torus. The inner and outer rings, drawn solid, correspond to the solidly drawn parts of the space curve in Figure 5.4 (a). The two dashed rings on the torus correspond to the dashed parts of the space curve. The solidly-drawn right branch and dashed left branch of the curve meet at the same point at infinity, P_∞. The right branch together with P_∞ forms a topological loop. Likewise for the left branch.

5.6. SINGULARITIES AND GENUS

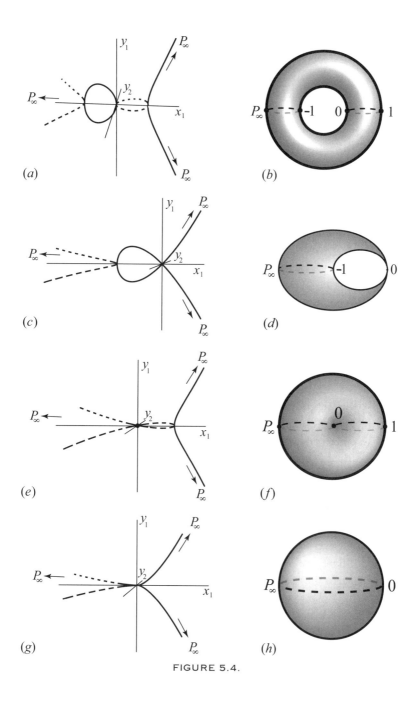

FIGURE 5.4.

Now let's see if we can trick the above curve into gaining a singularity while maintaining its genus. Our plan is to let the root $x = 1$ approach the root $x = 0$, which we keep fixed. That is, in $p = y^2 - (x + 1)x(x - \epsilon)$, let $\epsilon > 0$ approach zero. As this happens, the dotted loop in Figure 5.4 (*a*) gets smaller and smaller and in Figure 5.4 (*c*) we see that in the limit it has shrunk to a point — the origin — and in the (x_1, y_1)-plane the branch nose sharpens to a 45-degree angle as it touches the origin. We have created an alpha curve with a singularity (a node) at the origin. And the genus? Because the dotted loop collapsed to a point, the two solid loops of the torus in Figure 5.4 (*b*) now touch, the topological picture having morphed into a sphere with two horns touching at one point, shown in Figure 5.4 (*d*). But a sphere has genus 0, and those touching horns represent a 2-to-1 identification. From Definition 5.9, identifying finitely many points doesn't change the genus! So precisely when the singularity is created, the genus decreases.

Suppose that instead of letting the root $x = 1$ approach the origin, we let the root $x = -1$ do that. Now it's the solid loop in Figure 5.4 (*a*) that shrinks to a point. After this shrinking, there again appears an alpha curve, but this time it's in the (x_1, y_2)-plane, as shown in Figure 5.4 (*e*). Both partials of the new polynomial $y^2 - x^2(x - 1)$ vanish at the origin; the origin is a node. Figure 5.4 (*f*) shows that topologically we again get a sphere with two points identified to one. Instead of pulling out points to form touching horns, we essentially pinch the north and south poles together. But once again we see that the genus decreases just as a singularity is created.

What happens if *both* $x = 1$ and $x = -1$ approach $x = 0$? In that case both the solid and dashed loops in Figure 5.4 (*b*) shrink to a point. For example Figure 5.4 (*d*) could represent a half-way stage in this process, the morphing being completed by letting -1 approach 0. Then the solid inner loop shrinks towards 0 and the horned sphere becomes a sphere with no points identified. Or, the figure's third row could represent the half-way stage, the process finishing with the right dashed loop in Figure 5.4(*f*) shrinking to a point. That moves the two identified points to a point on the equator of what will be the final, ordinary sphere. Shrinking both loops in Figure 5.4 (*b*) to a point changes the space curve in Figure 5.4 (*a*) to two cusps — one in the usual (x_1, y_1)-plane, and the dashed one in the (x_1, y_2)-plane.

In all these examples of roots coalescing, at least one loop collapsed to a point. That is, all points in the loop became identified to one point. This is much stronger than simply identifying finitely many points to a point. In the first case, the genus decreases; in the second case, the identification is mild enough that the genus is unaffected.

5.6. SINGULARITIES AND GENUS

We will show in Example 5.16 starting on p. 114 that there exist curves of any genus by showing that the polynomial

$$y^2 - (x^2 - 1^2)(x^2 - 2^2) \cdots (x^2 - n^2)$$

defines a projective curve of genus $n-1$. We can coalesce roots of this polynomial to create various singularities, and in every case creating any kind of singularity reduces the genus. For any of these curves we can draw a real 1-dimensional curve in (x_1, y_1, y_2)-space as before, and we get branches and loops in the (x_1, y_1)-plane and in the (x_1, y_2)-plane. Parallel 3-spaces in $\mathbb{R}^4$ define real curves that fit together to form a surface.

For example, $n = 3$ gives $y^2 - (x^2 - 1^2)(x^2 - 2^2)(x^2 - 3^2)$, which defines a curve of genus 2. Figure 5.5 shows its graph in the real plane and the corresponding surface of genus 2.

Each of the four branch directions heads towards a common point P_∞, so that the two points at infinity marked on the surface are identified. This does not change the genus of 2. In the real-curve sketch, we can imagine dashed loops in the (x_1, y_2)-plane spanning the voids between the solidly drawn loops and branches in the picture.

Letting both roots $x = \pm 1$ coalesce to $x = 0$ pulls the solidly drawn loops together to form a figure 8, the origin becoming a node. Figure 5.6

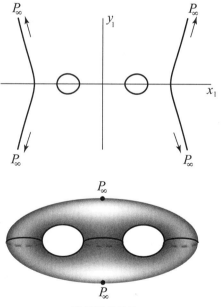

FIGURE 5.5.

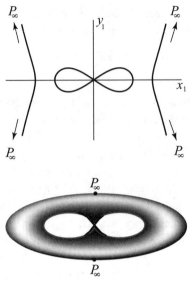

FIGURE 5.6.

shows the real part and the corresponding topological picture. The node forces the genus to decrease from 2 to 1. We can think of the topological surface as formed by starting with an ordinary torus and from opposite points on the inner waist, pulling points towards each other to form two little horns that grow until they touch at a point. The surface in Figure 5.6 can be thought of as formed from a torus with two vertically-oriented horns. Thus this surface has another 2-to-1 point identification in addition to the two identified points at infinity.

In Figure 5.7 we go a step further and let not only -2 coalesce to -1, but also 2 coalesce to 1. In this way we create *two* additional singularities — both nodes — and each collects its debt in that the genus decreases by one for each node created. The topological picture is a sphere with three 2-to-1 identifications, so its genus is 0.

Example 5.12. The principles in Chapter 1's "Designer Curves" section make it easy to create an irreducible curve with, say, one ordinary singularity of order 4 at the origin. The product $xy(x - y)(x + y)$ defines four lines through the origin. We may bound the real picture by adding higher even powers of x and y. One such curve in $\mathbb{P}^2(\mathbb{C})$ is defined by $p(x, y) = xy(x^2 - y^2) + x^6 + y^6$, whose part in $\mathbb{R}^2$ is depicted in Figure 5.8. It's easy to check that in $\mathbb{P}^2(\mathbb{C})$, the curve is everywhere smooth except at the origin. This curve has degree 6, and if it had no singularities

5.6. SINGULARITIES AND GENUS

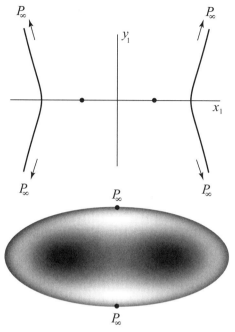

FIGURE 5.7.

FIGURE 5.8.

its genus would be $g = \frac{5 \cdot 4}{2} = 10$. But it has an ordinary singularity of order 4 similar to what we would have starting with 4 randomly selected lines (which create $\frac{4 \cdot 3}{2} = 6$ nodes) and moving the lines so the 6 nodes coalesce to form one ordinary singularity of order 4. Each node decreases the genus by one, so the actual genus of the topological surface of $C(p)$ is $10 - 6 = 4$.

We've encountered the formula $\frac{(n-1)(n-2)}{2}$ twice — once in the Genus Formula (Theorem 4.5 on p. 88), and again in Theorem 5.2 on p. 101, giving the maximum number of nodes of an irreducible curve of degree n. Is the same formula arising in apparently different contexts trying to tell us something? We've met examples in this section showing that as we reduce the genus of certain curves by squeezing a loop in the curve to a point, a singularity is introduced, and simultaneously with that the genus decreases. For example, look again at the process of squeezing down to the point 0 the loop between -1 and 0 in Figure 5.4(a) or (b) on p. 105. After squeezing, the part around 0 in picture (f) consists of two disks touching at just that point 0. This is exactly the topological picture around a node. We can reverse the "movie," expanding the point to a small loop to decrease the node count and increase the genus. So on the one hand, we have the picture of an irreducible nonsingular projective curve of degree n having genus $\frac{(n-1)(n-2)}{2}$. On the other hand, there's the picture of an irreducible nonsingular projective curve of degree n having $\frac{(n-1)(n-2)}{2}$ nodes, which we met on p. 101 just after Theorem 5.2. We can look at these as opposite extremes of a spectrum.

5.7 A More General Genus Formula

Examples in the last section suggest that for an irreducible curve C of degree n having only ordinary singularities, the genus $\frac{(n-1)(n-2)}{2}$ decreases by 1 if the curve's only singularity is a node. Also, by drawing together randomly-chosen lines, we argued that an ordinary singularity of order r can be thought of as $\frac{r(r-1)}{2}$ nodes piled up on each other at a common location, so that an ordinary singularity of order r decreases $\frac{(n-1)(n-2)}{2}$ by $\frac{r(r-1)}{2}$. Notice that $\frac{r(r-1)}{2}$ is 1 when $r = 2$, which corresponds to a single node. ($r = 1$ corresponds to a nonsingular point.) These decreases happen at each singularity, and the decreases add. This is expressed in this theorem generalizing Theorem 4.5 on p. 88.

Theorem 5.4. Let $C \subset \mathbb{P}^2(\mathbb{C})$ be an irreducible curve of degree n having only ordinary singularities, and suppose $r_1, \cdots, r_N$ are the orders of these singularities. Then the genus of C is

$$g = \frac{(n-1)(n-2)}{2} - \sum_{i=1}^{N} \frac{r_i(r_i-1)}{2}. \qquad (5.2)$$

For a proof, see [Walker], Chapter. VI, section 5.2. We can read off from this formula the topological structure of the curve: C is a real 2-manifold of genus given by (5.2) in which for each of the N distinct points P_i, r_i little horns rise from the surface, come together and meet at P_i. Since the genus cannot be negative, (5.2) shows in this case how the degree restricts the possible number of singularities.

5.8 NON-ORDINARY SINGULARITIES

Up to now we've focused on ordinary singularities. They are the best-behaved and the easiest to understand. Geometrically, if P is an ordinary singularity of order $r \geq 2$, then the part of $C(p)$ in some neighborhood of P is the union of r disks all mutually containing P and only P. The topological closure of any "disk minus P" is the graph of a function of one complex variable analytic at P. The complex tangent lines to the disks at P are distinct, any two intersecting in P. This makes it easy to devise all sorts of ordinary and non-ordinary singularities because the tangent structure to a curve $C(p)$ at $P = (x_0, y_0)$ is defined by the initial, or lowest-degree part of p when expanded about (x_0, y_0). (See p. 23.) For simplicity, we assume P is the origin unless stated otherwise.

Example 5.13. The alpha curve's polynomial $y^2 - x^2 - x^3$ has lowest-degree part $y^2 - x^2$, so the tangent space is given by $y^2 - x^2 = 0$ and therefore consists of the two lines $y = \pm x$. Because they are distinct, the singularity is ordinary.

The argument in Example 5.2 on p. 96 can be restated this way:

Example 5.14. The cusp $y^2 - x^3$ has lowest-degree part y^2, and $y^2 = 0$ defines a double line. The lines are not distinct, so the singularity is not ordinary.

The following is easy to prove.

Theorem 5.5. A singularity of $C(p)$ at the origin is ordinary if and only if all factors of the lowest-degree part of p are distinct.

We can make the origin a quite fancy non-ordinary singularity by taking the union of curves that share one or more tangent lines through the origin.

Example 5.15. Here are some polynomials each defining a non-ordinary singularity at the origin. It turns out that adding sufficiently high even powers of x and y makes each of these polynomials irreducible.

- $(y^2 - x^3)(y^2 - 2x^3)$ (Two tangent cusps)
- $(y^2 - x^3)(y^2 - x^5)$ (Cusps of different types)
- $y(y - x^2)(y - x^4)$
- $x(y^3 - x^5)$
- $(x^2 + y^2)^3 - (x^2 - y^2)^2$ (Four leaf rose)

Can non-ordinary singularities decrease genus more dramatically then ordinary ones? Theorem 5.4 on p. 110 tells us that each ordinary singularity of order r decreases the genus by $\frac{r(r-1)}{2}$. Non-ordinary singularities possess nondistinct tangent lines, and shared tangent lines can increase the node count. This happens in large part because tangency increases the "order of contact," or multiplicity of intersection. The parabolas $y = x^2$ and $y = -x^2$ illustrate this. Their union C has a singularity of order 2 at the origin. That is, 2 is the lowest multiplicity in which a line through the origin can intersect C, which is the same as the order of a node. If we push the upper parabola down a bit, the parabolas intersect in two nodes. As the upper parabola returns to its original position, the two nodes coalesce, so although the non-ordinary singularity has order 2, it is composed of 2 nodes, not 1. In this case, does a non-ordinary singularity act more powerfully to reduce genus? We can get a clue from looking again at the curve of genus 2 defined by $p = y^2 - (x^2 - 1^2)(x^2 - 2^2)(x^2 - 3^2)$. We saw that in letting the two roots ± 1 coalesce (both moving to $x = 0$), Figure 5.5 morphed into Figure 5.6. The figure 8 there correctly suggests that the created singularity is a node, and the genus correspondingly decreased by 1. Now let the two roots ± 2 *also* coalesce to the same point $x = 0$. That means the original polynomial morphs to $y^2 - x^4(x^2 - 3^2)$. Figure 5.9 shows the real part of the curve.

Near the origin, the real part resembles the two-parabola curve. Moving only the upper part downward a bit creates two nodes, suggesting that the singularity decreases the original curve's genus from 2 to 0. In fact, it does. The complex curve's intersection with (x_1, y_1, y_2)-space reveals two branches in the (x_1, y_2)-plane that meet at infinity and can be thought of as forming a great circle on a sphere passing though the north and south poles. The pinched 8 of Figure 5.9 corresponds to the equator. So the topology of this curve is a sphere with two 2-to-1 identifications: the two poles are identified and so are two opposite points on the equator. The non-ordinary singularity has reduced the original genus by 2, so the curve's genus is 0.

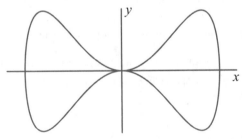

FIGURE 5.9.

5.8. NON-ORDINARY SINGULARITIES

To calculate how much a singularity reduces the genus, we need to know how many nodes comprise it. This important count of double points is often denoted by δ. For a non-ordinary singularity, a more powerful tool is needed to calculate δ. Here's how it works. If the origin is a singularity of the irreducible curve $C(p)$, find the intersection multiplicity at the singularity of the two "partial derivative curves" $C(p_x)$ and $C(p_y)$. We call this the *Milnor multiplicity* and denote it by μ. For example, in the two-parabolas example with $p = (y - x^2)(y + x^2) = y^2 - x^4$, we have $p_x = 4x^3$ and $p_y = 2y$. $C(p_x)$ is the "triple y-axis" and $C(p_y)$ is the ordinary x-axis. They intersect in multiplicity 3, so $\mu = 3$. Now μ and δ are related by the basic Milnor-Jung formula:

$$\delta = \frac{\mu + r - 1}{2}, \qquad (5.3)$$

where r is the number of branches through the singularity. In our case $r = 2$, so $\delta = \frac{3+2-1}{2} = 2$, which agrees with what our intuition suggested. For a discussion of these concepts, see [B-K], Chapter III, 8.5 (Topology of Singularities).

With two tangent cusps, it becomes harder to intuit the answer. For example, take two cusps given by $y^2 - x^3$ and $y^2 - 4x^3$. Their product, plus terms bounding the real picture, gives

$$p = (y^2 - x^3)(y^2 - 4x^3) + x^8 + y^8$$

which defines an irreducible curve $C(p)$ of degree 8. Figure 5.10 shows the real portion.

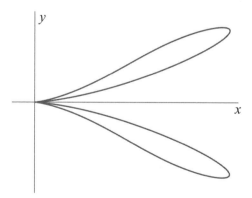

FIGURE 5.10.

It is straightforward to check that the origin is the only singularity of this curve in $\mathbb{P}^2(\mathbb{C})$. The partials are

$$p_x = x^2(24x^3 - 15y^2 + 8x^5), \quad p_y = y(4y^2 - 10x^3 + 8y^6).$$

The corresponding curves are $C(p_x) = C_1 \cup C_2$, where C_1 is the double line $C(x^2)$, and C_2 is $C(24x^3 - 15y^2 + 8x^5)$. $C(p_y) = C_1' \cup C_2'$, where C_1' is the line $C(y)$ and C_2' is $C(4y^2 - 10x^3 + 8y^6)$. The intersection multiplicity at the origin is the sum of the intersection multiplicities of each C_i with each C_j'. These are

- $m(C_1, C_1') = 2$
- $m(C_1, C_2') = 4$
- $m(C_2, C_1') = 3$
- $m(C_2, C_2') = 6$, found as the order at 0 of

$$\text{resultant}(24x^3 - 15y^2 + 8x^5, 4y^2 - 10x^3 + 8y^6, x).$$

These sum to $\mu = 15$. There are two cusps, so there are two branches through the origin. Therefore $r = 2$. The Milnor-Jung formula gives

$$\delta = \frac{15 + 2 - 1}{2} = 8.$$

If our degree 8 curve had no singularities, its genus would be $\frac{7 \cdot 6}{2} = 21$. But $\delta = 8$, which measures the effective number of nodes, reduces this to $21 - 8 = 13$.

We can now further improve Theorem 5.4 on p. 110:

Theorem 5.6. Let C be an irreducible curve of degree n having singularities at $P_1, \cdots, P_N$, and suppose $\delta_1, \cdots, \delta_N$ are double-point counts given in (5.3). Then the genus of C is

$$g = \frac{(n-1)(n-2)}{2} - \sum_{i=1}^{N} \delta_i. \tag{5.4}$$

Example 5.16. We stated on page 107 that there are curves of any genus. Specifically, the polynomial

$$p = y^2 - (x^2 - 1^2)(x^2 - 2^2) \cdots (x^2 - n^2) \tag{5.5}$$

defines a projective curve of genus $n - 1$. We argue as follows. First, the affine curve in the complex (x, y)-plane has no singularities — that is, there are no points on the curve where $p_x = p_y = 0$ — because $p_y = 2y^2$, so $y = 0$. From the form of (5.5), if $(x, 0)$ is on the curve, then x must be $\pm j$, $1 \leq j \leq n$. But the jth term of p_x is $2x$ in the jth place and $(x^2 - j^2)$ in all other places. Then $p_x(\pm i, 0)$ is nonzero at each j, because any kth term ($k \neq i$) of p_x is certainly zero, while the jth term is $\pm 2j$ times something nonzero. Therefore any singularity must be at infinity. Now, homogenize (5.5) with respect to z and dehomogenize at $y = 1$. This yields

$$q(x, z) = y^2 z^{2n-2} - (x^2 - z^2)(x^2 - 2^2 z^2) \cdots (x^2 - n^2 z^2). \quad (5.6)$$

Upon expanding this, we see that the lowest nonzero power of x is 2; likewise for z. Therefore $q_x(0, 0) = q_y(0, 0) = 0$, so there's a singularity at infinity. Now let's apply (5.3), the Milnor-Jung formula

$$\delta = \frac{\mu + r - 1}{2}.$$

The form of (5.5) means $r = 2$. Also, it is not hard to check that the order of q_x at $(0, 0)$ is $2n - 1$, and that the order of q_y at $(0, 0)$ is $2n - 3$, which leads to $\mu = (2n - 1)(2n - 3)$. Substituting this μ and r into the Milnor-Jung formula and simplifying gives $\delta = 2n^2 - 4n + 2$. Therefore our genus formula (5.4) yields

$$g = \frac{(2n - 1)(2n - 2)}{2} - (2n^2 - 4n + 2),$$

which simplifies to $g = n - 1$.

5.9 Further Examples

Curves of the Form $y^m = x^n$

An important source of nonordinary singularities are irreducible cusp curves $y^m = x^n$, with $n, m > 1$. Where are their singularities? How much do they reduce the genus of the curve? To get answers, let's start by noting that if $y^m - x^n$ is irreducible, then m and n must be relatively prime. An argument by contradiction is easy: if m and n shared a common divisor $k > 1$, then we could write $m = km'$ and $n = kn'$, which would mean $y^m - x^n = y^{km'} - x^{kn'}$. This last is divisible by $y^{m'} - x^{n'}$ and therefore is not irreducible.

We therefore assume m and n relatively prime, and we take $n > m$. Writing $p = y^m - x^n$, we see that since $p_x = -nx^{n-1}$ and $p_y = my^{m-1}$, these partials are both 0 only at the origin, so that's the only singularity of the cusp curve in the (x, y)-plane. Homogenizing p and dehomogenizing at $y = 1$ shows that the only other singularity is at the end of the y-axis. We can use the Milnor-Jung formula to count the number of equivalent nodes at each of these two singularities. At the origin of the (x, y)-plane, the simple form of the two partials tells us that μ there is $(n-1)(m-1)$. With $r = 1$, that means δ at the origin is $\frac{(n-1)(m-1)}{2}$. (Our assumptions on m and n imply that at least one of $n-1$, $m-1$ is even.) Homogenizing p and dehomogenizing at $y = 1$ gives $z^{n-m} - x^n$ from which we similarly find that δ there is $\frac{(n-1)(n-m-1)}{2}$. The sum of these two node counts is $\frac{(n-1)(n-2)}{2}$. This is the same as the genus of a nonsingular curve of degree n, so together the two singularities reduce any such cusp curve's genus to 0. We will meet a dramatic shortcut to this fact on p. 157.

AN EXAMPLE WITH REPEATED TANGENT LINES

The curve defined by $p = x^3y^4 + x^8 + y^8$, depicted in Figure 5.11, showcases the power of Milnor multiplicity. The curve is irreducible and the origin is its only singularity. It has $r = 2$ branches through the origin, one tangent to the x-axis, the other tangent to the y-axis.

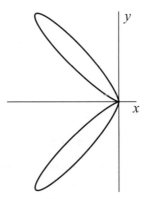

FIGURE 5.11.

Because the initial part x^3y^4 defines 4 copies of the x-axis and 3 copies of the y-axis, it is not obvious just how many nodes comprise it. The partials of p are $p_x = 3x^2y^4 + 8x^7$ and $p_y = 4x^3y^3 + 8y^7$. To find the multiplicity in which $C(p_x)$ and $C(p_y)$ intersect at the origin, take resultant(p_x, p_y, x);

its order in y turns out to be 41, so

$$\delta = \frac{\mu + r - 1}{2} = \frac{41 + 2 - 1}{2} = 21.$$

If this degree 8 curve were nonsingular, its genus would be $\frac{7 \cdot 6}{2} = 21$. With a node count of 21, this curve has genus 0.

5.10 Singularities versus Doing Math on Curves

Algebraic curves without singularities are especially pleasant because they can serve in a natural way as a domain for functions, allowing us to "do math" on them, much as one does calculus of several variables on differential manifolds or complex variable theory on the Riemann sphere. Can we transport to any projective curve elementary complex function theory in one variable? Singularities can present a problem in this regard, although we'll be able to overcome it. To illustrate the difficulty, let's compare what happens when we try to evaluate, say, $\frac{y}{x}$ at points of a simple nonsingular curve C_1 such as the complex x-axis, versus evaluating $\frac{y}{x}$ at points of the alpha curve $C_2 = C(y^2 - x^2(x+1))$, which has a node at the origin.

Example 5.17. At each point of the complex x-axis C_1, y is zero. Therefore $\frac{y}{x}$ is zero whenever $x \neq 0$. At $x = 0$, no matter how we approach $x = 0$ within the x-axis, the limit of the values is zero, so it is natural to define the value of $\frac{y}{x}$ to be zero there. In this way $\frac{y}{x}$ becomes single-valued at each point of C_1, thus making $\frac{y}{x}$ a function on the curve.

Example 5.18. Let's try the above argument on the alpha curve C_2. In evaluating $\frac{y}{x}$ at points of C_2, the only problematic point is the node at the origin, where both x and y are zero. There are two complex lines tangent to the curve there — the lines defined by the initial part $y^2 - x^2$ of $y^2 - x^2(x+1)$. Since $y^2 - x^2 = (y-x)(y+x)$, the complex tangent lines have slope $+1$ and -1. If we approach the origin along the curve's branch tangent to the line $y = x$, the values of x and y becomes more nearly equal to each other, so $\frac{y}{x}$ approaches the value $+1$. Approaching the origin along the other branch similarly gives a limiting value of -1. Assigning these limits at the origin gives two different values to $\frac{y}{x}$ there, so $\frac{y}{x}$ isn't a function on C_2.

What can we do? The problem at the node arose because at one point there were two limiting values for $\frac{y}{x}$. If we could somehow separate the two branches so we don't have different ones passing through the same point, that problem wouldn't arise. Is there some notion of equivalence between curves which is 1:1 between branches, but which allows their centers to separate into distinct points? None of the familiar candidates such as biholomorphism, diffeomorphism or topological transformation works because none of them splits up points. What sort of transformation can separate points so that different branches have different centers? For a little intuition, think of the alpha curve in the real setting as created from a long, springy wire deformed into an alpha shape. We could transform this shape by simply letting go, allowing the wire to return to its original straight shape. Doing this separates the two branches as well as their centers, and the straight line serves in a natural way as a curve on which to do mathematics. Another possible transformation is to hold one branch on a table top and lift the other branch upward to separate the branch centers. The curve then becomes a nonsingular space curve on which we can do mathematics.

Fortunately for our subject, there *is* a notion of equivalence that can perform the required magic. It is called *birational equivalence*, with birationally equivalent curves being connected through *birational transformations*. In the next sections we make the notion of "function on a curve" more precise, then define birational transformation and equivalence and give some examples. After that, we sketch how birational transformations can be used to desingularize algebraic curves, thus creating a canvas on which we can do elementary complex function theory.

5.11 The Function Field of an Irreducible Curve

Though we didn't explicitly say it above, at non-problematic points of C_1 or C_2 in Examples 5.17 and 5.18, we evaluated $\frac{y}{x}$ by simply substituting coordinates of points of the curve into $\frac{y}{x}$. The situation for polynomials is much nicer, because even for a curve $C \subset \mathbb{C}^2$ having singularities, we can always evaluate a polynomial $q \in \mathbb{C}[x, y]$ at each point of C by this method. This gives rise to a new phenomenon not encountered with polynomials on $\mathbb{R}^n$ or $\mathbb{C}^n$. On $\mathbb{R}^n$ or $\mathbb{C}^n$, if an n-variable polynomial is zero at each point, it's the 0-polynomial. But on the alpha curve $C(y^2 - x^2(x+1))$, there are infinitely many nonzero polynomials which are zero everywhere on the curve — they're all the nonzero multiples of the defining polynomial $y^2 - x^2(x + 1)$. There is a standard way to make the situation resemble that of $\mathbb{R}^n$ or $\mathbb{C}^n$, restoring the idea that there's just one function identically

5.12. BIRATIONAL EQUIVALENCE

zero on the curve. We look at this now.

Begin with a curve $C = C(p) \subset \mathbb{C}^2$, where p is irreducible in $\mathbb{C}[x, y]$. A polynomial $q \in \mathbb{C}[x, y]$, irreducible or not, defines a complex-valued function by restricting its domain to C. At each point of C, p itself is always zero, as is any multiple of p by a polynomial r. The set of all multiples of p forms an ideal (p) in $\mathbb{C}[x, y]$, and because p is irreducible, (p) is prime — that is, if a and b are in $\mathbb{C}[x, y] \setminus (p)$, then so is their product ab. This says that if neither a nor b has p as a factor, then their product doesn't either. In the quotient ring $\mathbb{C}[x, y]/(p)$, the ideal (p) plays the role of the zero-element in the additive group of the ring. Now (p) being prime says that $\mathbb{C}[x, y] \setminus (p)$ is closed under multiplication, so the quotient ring $\mathbb{C}[x, y]/(p)$ has no zero divisors. Therefore $\mathbb{C}[x, y]/(p)$ is an integral domain, and from an integral domain we can construct in the standard way its field of fractions. We denote this field by K_C. The only element of K_C identically zero on C is now the 0-element of K_C.

Definition 5.10. The field K_C constructed above for an irreducible curve $C \subset \mathbb{C}^2$ is called the *function field of* K_C.

The big question is, are the elements of K_C actually functions on C? If an element of K_C is q_1/q_2 with $q_i \in \mathbb{C}[x, y]/(p)$, then this quotient has a well-defined value at any point $P \in C$ at which not both q_1 and q_2 are zero. If q_2 is zero at P and q_1 is nonzero there, then we assign the value $\infty \in \mathbb{P}^1(\mathbb{C})$. The remaining question is, can we uniquely assign a value to q_1/q_2 if *both* q_1 and q_2 are zero at P? We will see that the answer is "maybe not" if C is singular at P and "definitely yes" if C is nonsingular at P. It's therefore important to appropriately desingularize a curve if we wish to do mathematics on it. The key to desingularization is *birational equivalence*, which we turn to now.

5.12 BIRATIONAL EQUIVALENCE

In this book, we will take "Doing math on an algebraic curve $C \subset \mathbb{P}^2(\mathbb{C})$" to mean doing elementary complex analysis in one variable on that curve. In a typical course on one complex variables, the functions are defined on the Riemann sphere $\mathbb{C} \cup \infty$ and take values there. The course is therefore complex variables on the projective algebraic curve $\mathbb{P}^1(\mathbb{C})$. We wish to show how the domain $\mathbb{P}^1(\mathbb{C})$ can be replaced by any projective algebraic curve. We begin with some basic ideas in this chapter, and develop them further in the next.

A basic prerequisite for doing complex variables on a curve C is that all elements of K_C should have well-defined values taken in $\mathbb{C} \cup \infty$. This will always hold if the curve is nonsingular. If function theories are to look alike on different curves, then the curves' function fields ought to be the same. It turns out that sharing the same set of functions is sufficient for algebraic curves: if curves C and C' have isomorphic function fields K_C and $K_{C'}$, then the cluster of facts associated with C is essentially identical to those for C'. Let's make a formal definition.

Definition 5.11. Function fields K_C and $K_{C'}$ are $\mathbb{C}$-isomorphic if there is a 1:1 onto correspondence between their elements that is the identity on $\mathbb{C}$ and that preserves all field operations.

Definition 5.12. If the function fields K_C and $K_{C'}$ of curves C and C' are $\mathbb{C}$-isomorphic, then we say C and C' are *birationally equivalent*.

You may wonder, "What's *birational* about the equivalence?" Since a field is a collection of ratios, under a field isomorphism any element or ratio in one field maps into a ratio in the other, and vice-versa. In this sense, isomorphic fields could be called birationally equivalent fields. "But," you may counter, "we're saying that *curves* are birationally equivalent, not fields." The answer is that the function field of an irreducible algebraic curve generates a corresponding curve in a natural way. Here's the idea.

First, the function field of any irreducible curve can be written as $\mathbb{C}(x, y)$, where one of x, y is an indeterminate element and the other is algebraically dependent on the indeterminant. For example, suppose our curve is $C(p)$, with $p(x, y)$ irreducible. If both x and y appear in $p(x, y)$, then y may be regarded as algebraically dependent on the indeterminate x. That is, y satisfies a polynomial equation in y with coefficients in $\mathbb{C}[x]$. By looking at the coefficients as coming from the field $\mathbb{C}(x)$, $\mathbb{C}(x, y)$ becomes an algebraic field extension of the field $\mathbb{C}(x)$ in one indeterminate. The roles of x and y can be reversed, with x regarded as algebraically dependent on the indeterminate y, with x satisfying a polynomial equation in x with coefficients in $\mathbb{C}[y]$ or $\mathbb{C}(y)$. The field $\mathbb{C}(x, y)$ is then an algebraic extension of $\mathbb{C}(y)$. We will usually take x as the indeterminate and y as dependent on x. If not both x and y appear in $p(x, y)$, then since p is irreducible, we can assume $p(x, y)$ has the form $x - c$ or $y - c$ and the fields are simply $\mathbb{C}(x)$ or $\mathbb{C}(y)$ rather than proper extensions of them.

We said above that the function field of an irreducible algebraic curve generates a corresponding curve in a natural way. Here's how. Generators

x, y of the function field $K_C = \mathbb{C}(x, y)$, although not uniquely determined, can be used to trace out a curve C': a complex value x_0 of x determines finitely many y-values y_i determined by the algebraic dependence $p(x_0, y) = 0$, and as the values of x_0 fill out $\mathbb{C}$, the associated pairs (x_0, y_i) fill out C'. The ordered pair (x, y) is sometimes called a "generic point" of C'.

Since the above C and C' have the same function field, the curves are birationally equivalent. Let ϕ be a $\mathbb{C}$-isomorphism from K_C to $K_{C'}$. Each element of $\mathbb{C}(x, y)$ is rational in x and y, and each element in $\mathbb{C}(x', y')$ is rational in x' and y'. Therefore under the isomorphism, each element of one field corresponds to a rational expression in the other. In symbols, $\phi(x) = r(x', y')$ and $\phi(y) = s(x', y')$, where $r(x', y')$ and $s(x', y')$ are rational in x' and y'. These can be written more compactly as

$$\phi((x, y)) = (r(x', y'), s(x', y')).$$

Since ϕ^{-1} is an isomorphism from $K_{C'}$ to K_C, we have, similarly,

$$\phi^{-1}((x', y')) = (r'(x, y), s'(x, y)).$$

where r' and s' are rational in x and y.

The above discussion leads to a basic prescription: starting with a "bad" curve — one with singularities — we want to find a nonsingular curve birationally equivalent to it, for it is on nonsingular curves that evaluating functions is problem-free and doing one-variable complex function theory on such a curve runs smoothly. We say that any irreducible curve is a *model* of its function field. Our overall aim is to find a nonsingular model of a function field of any curve with singularities. Doing this, by whatever means, is known as *desingularizing* the curve. In the next section we illustrate doing this for some simple examples.

5.13 Examples of Birational Equivalence

It's time to illustrate the above ideas with specific examples. On p. 118 we intuitively imagined the real alpha curve as a long, springy wire deformed into an alpha shape, and we noted that we could transform this shape by simply letting go, allowing the wire to return to its original straight shape. This line serves as a natural domain for fully evaluatable rational functions. In Example 5.19 next, we show that in the complex setting, the line and the alpha curve are birationally equivalent. That means $\mathbb{C}$ is a model of the alpha curve's function field, and that "functions" on the alpha curve — elements of the curve's function field — can be transferred to $\mathbb{C}$ via function-field isomorphism.

Example 5.19. Our alpha curve has function field $K_C = \mathbb{C}(x, y)$, where y depends on x through $y^2 = x^2(x + 1)$. The function field of a line $C' = \mathbb{C}$ consists of the rational functions on the line, so is $K_{C'}$ is $\mathbb{C}(x')$. To make notation a little simpler, we write t instead of x'. We will establish that the line and alpha curve are birationally equivalent, which means that the complex line is a nonsingular model of the alpha curve.

To show that the line and alpha curve are birationally equivalent, we find an isomorphism $\phi : \mathbb{C}(t) \to \mathbb{C}(x, y)$. We will first determine the image of $t \in \mathbb{C}(t)$ under the map. From that, it will be easy to see which elements of $\mathbb{C}(t)$ map to x and y in $C(x, y)$. This is equivalent to getting a rational parametrization of x and y in terms of t, say $\{x = r(t), y = s(t)\}$. This parametrization establishes a link between equivalence of fields and equivalence of curves: as t fills out the complex line $\mathbb{C}$, the images under ϕ fill out the alpha curve in $\mathbb{C}^2$. One could more informatively call $\{x = r(t), y = s(t)\}$ a *birational* parametrization or map, establishing a birational equivalence between the line and the alpha curve. For convenience, we abuse notation a bit and also denote by ϕ the birational map from the line to the alpha curve.

Bézout's Theorem can be used to get an actual parametrization. The polynomial $y^2 - x^2(x + 1)$ defining the alpha curve has degree 3. Thus for $t \neq \pm 1$, any line $y = tx$ ($t \in \mathbb{C}$) intersects the node in multiplicity 2, and therefore any such line intersects the alpha curve in just one other point for a total of $3 \cdot 1$ points. Here $t \in \mathbb{C}$ represents the slope of a line through the origin, and the line intersects the alpha curve in one other point (x, y). In this way, Bézout's theorem provides a link between t and points of the curve. It's easy to determine where the line $y = tx$ intersects the curve $y^2 - x^2(x + 1)$, because substituting tx for y in $y^2 - x^2(x + 1)$ gives $t^2 x^2 = x^2(x + 1)$. At the non-node point of intersection, x is nonzero. We may therefore divide by x^2, giving $t^2 = x + 1$ — that is, $x = t^2 - 1$. Because $y = tx$, we obtain the parametrization

$$\{x = t^2 - 1, \quad y = t(t^2 - 1)\}. \tag{5.7}$$

Notice that (5.7) works even for $t = \pm 1$. Figure 5.12 illustrates this in the real setting.

5.13. EXAMPLES OF BIRATIONAL EQUIVALENCE

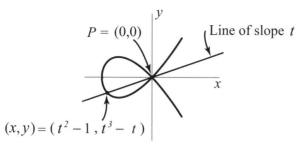

FIGURE 5.12.

This parametrization shows that the alpha curve's function field $\mathbb{C}(t^2 - 1, t(t^2 - 1))$ is $\mathbb{C}(t)$ since the field contains the quotient

$$\frac{y}{x} = \frac{t(t^2 - 1)}{t^2 - 1} = t. \quad (5.8)$$

Importantly, note that **the field isomorphism does not imply a 1:1 correspondence between points of the curves**. Figure 5.13 illustrates this in the real setting.

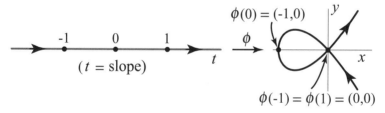

FIGURE 5.13.

The correspondences at the algebraic and geometric levels are illustrated in the top and bottom parts of Figure 5.14.

In transporting complex functions from the line to the alpha curve, note how (5.7) and (5.8) make the connection explicit. The simplest non-constant function on the curve $\mathbb{C}$ is t. What function on the alpha curve does t transport to? The answer appears in (5.8): it is $\frac{y}{x}$. The parametrization in (5.7) "paints" $\mathbb{C}$ onto the alpha curve, covering its node twice, and via (5.8) it shows that any value $t_0 \in \mathbb{C}$ assigned to t maps to the very same numeric value for $\frac{y}{x}$ at the corresponding point (x_0, y_0) in the alpha curve:

$$\frac{y_0}{x_0} = \frac{t_0(t_0^2 - 1)}{t_0^2 - 1} = t_0.$$

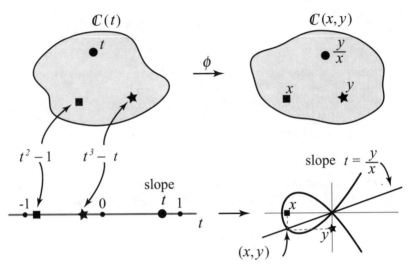

FIGURE 5.14.

In fact, we can extend t_0 to include ∞, because when t_0 has the value ∞, so does $\frac{y_0}{x_0}$.

Example 5.20. Example 5.19 illustrates how a rational t-parametrization of one curve leads to birational equivalence with $\mathbb{C}$. Being equivalent to $\mathbb{C}$ is a nice state of affairs since $\mathbb{C}$ is such a simple curve. What about the circle $x^2 + y^2 = 1$? Does it have a rational t-parametrization? Although its familiar parametrization $\{x = \cos(t), y = \sin(t)\}$ might suggest otherwise, it turns out that the idea used with the real alpha curve — we might call it the "rotating line approach" — also works here. Though any point on the circle will do, computation is simplified by choosing $P = (-1, 0)$. As with the alpha curve, we take lines through P of slope t, illustrated in Figure 5.15.

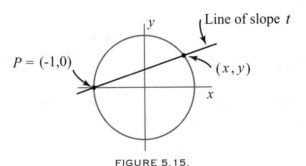

FIGURE 5.15.

5.13. EXAMPLES OF BIRATIONAL EQUIVALENCE

The line through P of slope t has equation $y = t(x + 1)$. Substitute this into $x^2 + y^2 = 1$ to get $x^2 + t^2(x + 1)^2 = 1$, which is quadratic in x:

$$(t^2 + 1)x^2 + 2t^2 x + (t^2 - 1) = 0.$$

The quadratic formula yields $x = (-t^2 \pm 1)/(t^2 + 1)$ for the x-coordinates of the two points of intersection. The -1 term in the numerator gives the point P, while the $+1$ term yields the variable point (x, y). We end up with the parametrization

$$\left\{ x = \frac{1 - t^2}{1 + t^2}, \quad y = \frac{2t}{1 + t^2} \right\}.$$

As $t \to \pm\infty$, the parametrized point tends to P. The parametrization also lets us see the circle's function field $\mathbb{C}(x, y)$ in very simple terms: t is the slope of the line $y = t(x + 1)$, and we can write that slope as $t = \frac{y}{x+1}$. But $\mathbb{C}(x, y)$ contains the quotient $\frac{y}{x+1}$, so $\mathbb{C}(x, y)$ is isomorphic to $\mathbb{C}(t)$.

On the surface of it, the rational parametrization of a circle seems to have little to do with the "transcendental" one, $\{x = \cos(t), y = \sin(t)\}$. However, if we replace the slope t in the rational parametrization by $\tan \theta$ and substitute this into the rational parametrization, we get

$$x = \frac{1 - \tan^2 \theta}{1 + \tan^2 \theta}, \quad y = \frac{2 \tan \theta}{1 + \tan^2 \theta}.$$

The expression for x can be rewritten as

$$x = \cos^2 \theta (1 - \tan^2 \theta) = \cos^2 \theta - \sin^2 \theta = \cos 2\theta,$$

while y simplifies to

$$y = 2 \frac{\sin \theta}{\cos \theta} \cos^2 \theta = 2 \sin \theta \cos \theta = \sin 2\theta.$$

Figure 5.16 shows that the only real difference between the parametrizations is the location of the angle's vertex.

The rational parametrization

$$\left\{ x = \frac{1 - t^2}{1 + t^2}, \quad y = \frac{2t}{1 + t^2} \right\}$$

of the unit circle $x^2 + y^2 = 1$ leads to a birational equivalence between the parameter line $\mathbb{C}$ and the circle, because the mapping $\psi : \mathbb{C}(x, y) \to \mathbb{C}(t)$ defined by

$$\psi(x) = \frac{1 - t^2}{1 + t^2}, \quad \psi(y) = \frac{2t}{1 + t^2}$$

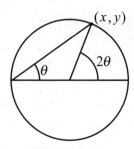

FIGURE 5.16.

is an isomorphism. The argument is similar to that used in Example 5.19. The parameter line and the circle are both nonsingular, so ψ defines a 1:1 onto map between their points.

Example 5.21. The mappings in the above two examples can be combined to get a birational equivalence between the unit circle C and the alpha curve C'. From Figure 5.15, we see that since the parameter t is the slope of the line through $(-1, 0)$ and (x, y), the point (x, y) on the circle determines $t = \frac{y}{x+1}$. By substituting, this in turn determines a point (x', y') on the alpha curve with parametrization $\{x' = t^2 - 1, \ y' = t^3 - t)\}$. We get

$$x' = \left(\frac{y}{x+1}\right)^2 - 1,$$

$$y' = \left(\frac{y}{x+1}\right)^3 - \frac{y}{x+1}.$$

In the real setting, Figure 5.17 shows that as a point on the circle starts in

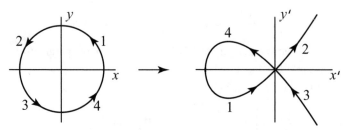

FIGURE 5.17.

the first quadrant and moves counterclockwise, the birational image in the alpha curve starts in the third quadrant and sweeps along the curve into the first, fourth and second quadrants, then cycles back into the third. The four points in which the circle intersects an axis map to points on the alpha curve as follows:

$$(1, 0) \longrightarrow (-1, 0)$$

$$(0, 1) \longrightarrow (0, 0)$$

$$(-1, 0) \longrightarrow \text{point at infinity}$$

$$(0, -1) \longrightarrow (0, 0)$$

Both $(0, 1)$ and $(0, -1)$ map to the node of the alpha curve.

In Example 5.19 starting on p. 122, the two points $t = 1$ and $t = -1$ in the parameter line each map to the origin of the alpha curve. In general there will be finitely many such exceptions. By introducing the concept of "place," those exceptions are removed. A place is a "germ of an analytic branch through a point." More informally, and sufficient for our purposes, a *place with center* P can be represented by the part of a branch through a point P within an open neighborhood small enough so that any two such representatives of different branches through P intersect in only P, and representatives of branches through different points don't intersect at all. The name makes sense: "place" is a local concept, but includes more than just a point. The following fact suggests the central importance of places; we state it here without proof (see [Fischer], Theorem 9.3, p. 169).

Theorem 5.7. Suppose C_1, C_2 are two birationally equivalent projective algebraic curves, and let ϕ denote a $\mathbb{C}$-isomorphism between their function fields: $\phi : K_{C_1} \to K_{C_2}$. Then ϕ induces a 1:1 onto correspondence between the places of C_1 and those of C_2.

In the next chapter, we say more about just how ϕ induces this correspondence. If C and C' are not only birationally equivalent, but are also nonsingular, then there's just one place through each point of C and through each point of C'. In that case there is a 1:1 onto correspondence between the points of C and C'. We discuss places further on pp. 154–155.

5.14 SPACE-CURVE MODELS

Why space curves? On p. 118, we imagined desingularizing the real alpha curve by keeping one of the node's two crossing wires on a table top and vertically lifting the other wire to separate the branches, resulting in a curve in $\mathbb{R}^3$. But a line desingularizes the alpha curve even more simply than the space curve, so why bother? The answer is that a plane curve can have millions of bad singularities, and there may exist no desingularization of the curve *in the plane*. Space curves become a necessity.

Definition 5.13. A *space curve in* $\mathbb{C}^n$ is the common set C of zeros of a collection of polynomials in n variables such that C has complex dimension 1 at each of its points P. Topologically, the part of C in some sufficiently small neighborhood of P consists of finitely many disks intersecting in only P.

In mathematics, the need for increasing the size of an object's living space is not an isolated phenomenon. For a little perspective, visualize a loose knot in $\mathbb{R}^3$. No matter how you orient the knot, its shadow on the ground must cross or double over on itself. Give it an extra dimension, however, and the shadow can be lifted to a knot having no self-intersections. Algebraic curves respond in a similar way when given one extra dimension of living space. It turns out that any algebraic curve in complex 2-space with a multitude of even the worst kinds of singularities is itself a projection or shadow of some nonsingular curve in complex 3-space. This is reminiscent of what happens with closed 2-manifolds in real 3-space. There exists a huge variety of these, including familiar examples such as the Klein bottle or the real projective plane, that cannot be represented in $\mathbb{R}^3$ without self-intersections or without identifying points as we do in making a Möbius strip from a rectangle. But in the one higher dimension of $\mathbb{R}^4$, they can happily exist without self-intersections or identifications. With algebraic curves, that one extra complex dimension makes all the difference.

One trick used earlier — the rotating line method — can be adapted to increase a singular curve's living space and construct a nonsingular curve having the same function field. We used this method to get a birational parametrization of the alpha curve, and we did the same for the circle. The idea also works beautifully to desingularize one or more ordinary singularities of arbitrarily high order. Although the phrase "rotating line" makes good geometric sense in the real setting, we abuse terminology a little and continue to use this terminology in the complex setting, too. Seeing how this method works for the alpha curve reveals the essential idea in its simplest form. After that, applying the method to a single higher order ordinary singularity and then to several such singularities allows the approach to unfold naturally. Therefore in Example 5.22 next, we use the rotating line method to make precise the intuitive idea of separating crossed wires, mentioned on p. 118.

5.14. Space-Curve Models

Example 5.22. In Example 5.19 on p. 122, we learned that the line of slope t through the origin of the alpha curve $y^2 = x^2(x+1)$ intersects the curve in the point $(t^2-1, \, t(t^2-1))$. Since under the parametrization the node corresponds to the two different values $t = \pm 1$, we can desingularize simply by using t as a tag in an additional, third coordinate. The function of this tag is that it lifts the alpha curve into a new curve C' in 3-space parametrized by
$$(t^2-1, \, t(t^2-1), \, t).$$
Notice that suppressing the third coordinate in this parametrization corresponds to projecting C' back down to C, making C the shadow of C'. Figure 5.18 shows this shadow — the original singular curve C in the (x, y)-plane — together with the desingularized curve C' that lies on the surface $t = \frac{y}{x}$. This real surface was sketched by drawing lines in the (x, y)-plane through the origin and then translating each line up or down by an amount equal to its slope.

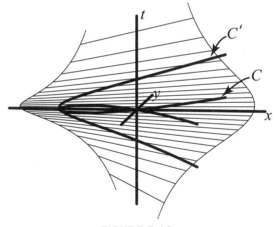

FIGURE 5.18.

Is C' algebraic in the sense that it is defined as the common set of zeros of polynomials in $\mathbb{C}[x, y, t]$? It is the intersection of two surfaces in 3-space. The equation for the first is obtained by omitting the first coordinate in the parametrization of C' to get the cubic cylinder $y = t^3 - t$. Omitting instead the second coordinate yields the parabolic cylinder $x = t^2 - 1$. Their intersection is C'. Figure 5.19 depicts this alternate way of visualizing C'. Figures 5.18 and 5.19 provide suggestive pictures of the situation, but everything actually takes place in the complex setting, where the curves and surfaces live in $\mathbb{C}^2$ and have real dimensions 2 and 4.

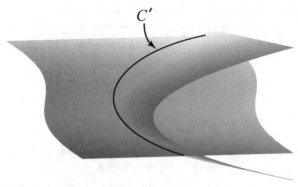

FIGURE 5.19.

The remaining question: *is C' nonsingular?* Yes, because the part of C' about any point of the curve is locally the graph of a function that's not only holomorphic, but polynomial. This comes from using t as a tag in the z-variable, and it means that we can choose the z-axis as domain. For example, the origin of C is a node, but the part of C' around the origin in $\mathbb{C}^3$ is the graph of the vector-polynomial function $(z^2 - 1, z^3 - z)$ and is therefore smooth. The vector polynomial tells us that all points of C' in $\mathbb{C}^3$ are nonsingular.

5.15 Resolving a Higher-Order Ordinary Singularity

The rotating line method is powerful enough to resolve any ordinary singularity of a curve C in $\mathbb{C}^2$ defined by an irreducible polynomial $p(x, y)$. We assume that the degree of p is n, and that coordinates are chosen so that the singularity is the origin.

Let L_t denote the complex line $y = tx$ through the origin with complex slope t, and let S_t denote the intersection of L_t with $C(p) \setminus \{(0, 0)\}$. If the singularity has order r, then by Bézout's theorem S_t consists of $n-r$ points (less any others that happen to be at infinity). For each value t_0 of t, lift S_{t_0} to the plane $\mathbb{C}^2$ in $\mathbb{C}^3 = \mathbb{C}_x \times \mathbb{C}_y \times \mathbb{C}_t$ defined by $t = t_0$. Geometrically, this lifts the intersection points from their original plane into 3-space so that the "complex lifting height" is the slope of the rotated line L_t. L_t always intersects the curve in the origin, but because the intersection of L_t is with $C(p) \setminus \{(0, 0)\}$, these intersection points are missing in the lifted curve. For example the lifted alpha curve would have two holes in it — one at

$(0, 0, 1)$ and one at $(0, 0, -1)$. We now plug up these holes by taking the topological closure of the lifted curve to get an algebraic space curve C'. Since the singularity is ordinary, the r branches of C through the origin have r distinct slopes, so each is lifted to a different height. In taking the topological closure, r different holes are plugged up. All this formalizes and extends the notion of separating crossed wires in the springy wire analog mentioned earlier. All branches have been separated and the resulting curve is nonsingular.

In mathematics, an object can often be constructed in two "dual" ways: building up by taking the union of smaller pieces, or cutting down by intersecting larger pieces. The above description uses the union approach: the space curve is the union of lifted sets of points. The intersection approach runs as follows. In $\mathbb{C}^3 = \mathbb{C}_x \times \mathbb{C}_y \times \mathbb{C}_t$, the zero set of $p(x, y)$ defines a surface of complex dimension two — an "algebraic variety." The real portion of this could be called the cylindrization in the t-direction of the original plane curve. Let V denote this surface minus the t-axis. In $\mathbb{C}_x \times \mathbb{C}_y \times \mathbb{C}_t$, the zero set of $y - tx$ defines another surface W. The real portion of this looks like a corkscrew making just one turn as $|t|$ increases without bound. The space curve is the topological closure of $V \cap W$. For further details, see [Fulton], Chapter 7.

5.16 Examples of Resolving an Ordinary Singularity

In this section we look at some concrete examples of desingularizing an ordinary singularity. Such singularities arise in many familiar examples of plane algebraic curves, but one good source are roses. Although they are typically defined in the real setting by polar coordinates, they are often algebraic, and a real rose defined by a polynomial $p(x, y)$ extends to an algebraic curve in $\mathbb{C}^2$ or $\mathbb{P}^2(\mathbb{C})$ when x and y take on complex values.

Example 5.23. Perhaps the simplest real rose is Bernoulli's lemniscate, which looks like a figure ∞. The principles outlined in section 1.10 on Designer Curves certainly allow us to create a two-leaved rose. With a little luck, our equation may turn out to be that of Bernoulli's lemniscate. Put the singularity at the origin, and let the two tangent lines to the figure ∞ be $y = \pm x$. We therefore make $x^2 - y^2$ the initial part, and choose some sufficiently large power $(x^2 + y^2)^n$ as leading term to bound the real curve.

Choosing $n = 2$ gives us this fourth-degree equation:

$$(x^2 + y^2)^2 = x^2 - y^2.$$

This in fact *is* the cartesian equation of the Bernoulli lemniscate.

To find a parametrization of this curve, let's apply the rotating line method. When $t^2 \neq 1$, the line $y = tx$ through the origin isn't tangent to the lemniscate and therefore intersects it at the origin with multiplicity 2. By Bézout's theorem, such a line intersects the curve in two other points (real, complex or at infinity) to make a total of 4 points. These two points are symmetric with respect to the origin, and in $\mathbb{R}^2$ we see them only when the line's slope t is in the real interval $(-1, 1)$. Figure 5.20 illustrates the situation.

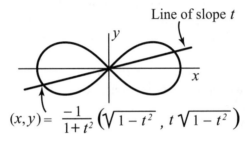

FIGURE 5.20.

Because the lemniscate's equation is quite simple, we can substitute tx for y in its equation and solve for nonzero x, getting

$$x = \pm \frac{\sqrt{1-t^2}}{1+t^2}.$$

Of course since $y = tx$, y is just t times this. The set S_t thus consists of the two points

$$\pm \left(\frac{\sqrt{1-t^2}}{1+t^2}, t\frac{\sqrt{1-t^2}}{1+t^2} \right),$$

and S_t lifts to

$$\left(\pm\frac{\sqrt{1-t^2}}{1+t^2}, \pm t\frac{\sqrt{1-t^2}}{1+t^2}, t \right).$$

The union of these liftings fills out a curve with points missing at $(0, 0, 1)$ and $(0, 0, -1)$. In the real portion, plugging these holes connects two bent

5.16. EXAMPLES OF RESOLVING AN ORDINARY SINGULARITY 133

semicircles, making a topological circle. Using Mathematica or Maple, plotting these points for real t gives a good sense of what the real space curve looks like, since we can rotate its plot on-screen. A simple physical experiment also approximates the real curve: give a thin rubber band a 90° twist and a pinch to form a figure 8, then slowly separate the touching points into a twisted circle. The circle should resemble what you get by twisting in opposite directions two antipodal points of a springy wire circle.

Example 5.24. We state without proof a few methods for creating a variety of roses, all of which have a single singularity at the origin. Fortunately, expressions involving $\Re(z^n)$ or $\Im(z^n)$ efficiently produce polynomials homogeneous in real variables x and y appearing in the Cartesian equations of many real roses. If we let x and y assume complex values, any such equation defines an algebraic curve in $\mathbb{C}^2$ whose real portion agrees with the original real curve.

- **Roses $r = \cos n\theta$ when n is odd.** For odd n, $r = \cos n\theta$ defines an n-leaved rose in $\mathbb{R}^2$, where we use the methods of analytic geometry to get the real sketch. In Cartesian coordinates, the polynomial equation for this real curve is

$$(x^2 + y^2)^{\frac{n+1}{2}} = \Re(z^n).$$

It has degree $n + 1$, and by letting x and y take complex values, it defines an affine or projective complex curve. It turns out that $(x^2 + y^2)^{\frac{n+1}{2}} - \Re(z^n)$ is irreducible, so the curve is, too. In $\mathbb{R}^2$, the rose has n distinct real tangent lines at the origin, and this correctly suggests that the extended curve in $\mathbb{C}^2$ has an ordinary singularity at the origin of order n. For odd n, the four equations

$$(x^2 + y^2)^{\frac{n+1}{2}} = \Re(z^n)$$
$$(x^2 + y^2)^{\frac{n+1}{2}} = \Im(z^n)$$
$$(x^2 + y^2)^{\frac{n+1}{2}} = -\Re(z^n)$$
$$(x^2 + y^2)^{\frac{n+1}{2}} = -\Im(z^n)$$

define four different real n-leaved roses successively rotated about $(0,0)$ by $\frac{\pi}{2n}$ (that is, by a quarter the petal angle $\frac{2\pi}{n}$), and these equations turn out to define irreducible algebraic curves in $\mathbb{C}^2$ when x and y assume complex values.

- **Roses $r = \cos n\theta$ when n is even.** When n is even, $r = \cos(n\theta)$ defines in $\mathbb{R}^2$ a real $2n$-petal rose, and we see n double-line tangents. That means the singularity at the origin in $\mathbb{C}^2$ must be nonordinary. The curve's Cartesian equation is $(x^2 + y^2)^{n+1} = \left(\Re(z^n)\right)^2$. It happens that the polynomial $(x^2 + y^2)^{n+1} - \left(\Re(z^n)\right)^2$ is irreducible, so the curve in $\mathbb{C}^2$ defined by it is irreducible. Interestingly, if n is instead odd, $(x^2 + y^2)^{n+1} - \left(\Re(z^n)\right)^2$ has the form $a^2 - b^2$ and is therefore reducible. Its curve is the union of two n-leaved roses, each the reflection about the origin of the other.

When n is even, replacing $\left(\Re(z^n)\right)^2$ by $\Re(z^{2n})$ splits up each double tangent to make the singularity ordinary. The equation for such a $2n$-leaved rose is therefore

$$(x^2 + y^2)^{n+1} = \Re(z^{2n}).$$

It can be checked that for any n — even or odd — the equations

$$(x^2 + y^2)^{n+1} = \Re(z^{2n})$$
$$(x^2 + y^2)^{n+1} = \Im(z^{2n})$$
$$(x^2 + y^2)^{n+1} = -\Re(z^{2n})$$
$$(x^2 + y^2)^{n+1} = -\Im(z^{2n})$$

define four different real roses of $2n$-petals, successively rotated about the origin by a quarter of the petal angle. In each case the corresponding curve in $\mathbb{C}^2$ is irreducible and has an ordinary singularity of order $2n$ at the origin. Note that $n = 1$ yields Bernoulli's lemniscate together with three other successive rotations of it by $45°$.

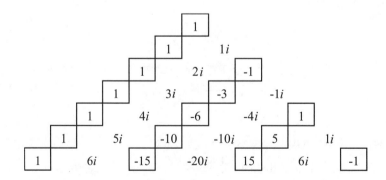

FIGURE 5.21.

Figure 5.21 facilitates converting $\Re$- and $\Im$-expressions into polynomials. Row n of the Pascal Triangle in the figure corresponds to the expansion

5.16. Examples of Resolving an Ordinary Singularity

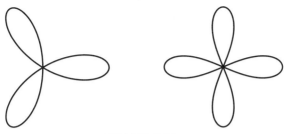

FIGURE 5.22.

of $(1+i)^n$. (The top "row" consisting of just 1, is row 0.) In row n, the coefficients of $\Re(z^n)$ have squares drawn around them. The coefficients of i in row n are those of $\Im(z^n)$.

Example 5.25. Choosing $n = 3$ in $(x^2 + y^2)^{\frac{n+1}{2}} = \Re(z^n)$ produces the equation
$$(x^2 + y^2)^2 = [\Re(z^3)] = x^3 - 3xy^2$$
whose real portion is a trefoil. This appears at the left in Figure 5.22. Choosing $n = 2$ in $(x^2 + y^2)^{n+1} = \Re(z^{2n})$ gives
$$(x^2 + y^2)^3 = \Re(z^4) = x^4 - 6x^2y^2 + y^4$$
whose real portion is the four-leaf rose, appearing on the right in the figure. All four tangent lines are distinct, so the singularity of the curve in $\mathbb{C}^2$ is ordinary. Compare this with the four-leaf rose appearing in Figure 5.1 on p. 96. That rose has a non-ordinary singularity at $(0, 0)$, and its equation is
$$(x^2 + y^2)^3 = [\Re(z^2)]^2 = (x^2 - y^2)^2 .$$

Example 5.26. We can use the rotating line method to desingularize the trefoil in Figure 5.22 defined by
$$p(x, y) = (x^2 + y^2)^2 - x^3 + 3xy^2.$$
This fourth-degree polynomial has order 3 at the origin, so its singularity is of order 3. The lowest-order part of p factors into $x(\sqrt{3}y+x)(\sqrt{3}y-x)$ so, as Figure 5.22 suggests, the three tangent lines at the origin are distinct and the singularity is ordinary. By Bézout's theorem, any line through the origin intersects the trefoil in $\deg(p) = 4$ points, and except for intersections with the three tangent lines, exactly 3 of these 4 are at the origin. Any non-tangent line through the origin uniquely determines a non-origin point of the trefoil in $\mathbb{C}^2$. Because one tangent line is the y-axis, we parametrize the

lines by $x = ty$ to avoid the case when $t = \infty$. Substituting ty for x in $p(x, y) = 0$ leads to

$$y^4(1 + t^2)^2 = (ty)^3 - 3ty^3,$$

which can be solved for nonzero y as $y = \frac{t^3-3t}{(1+t^2)^2}$. Put t in the third coordinate; the coordinates of the lifted S_t are then

$$\left(\frac{t^4 - 3t^2}{(1+t^2)^2}, \frac{t^3 - 3t}{(1+t^2)^2}, t \right).$$

As t varies throughout $\mathbb{C}$, these liftings fill out a curve with holes at $(0, 0, 0)$, $(0, 0, \sqrt{3})$, $(0, 0, -\sqrt{3})$. Plugging the holes by taking the topological closure produces a nonsingular space curve birationally equivalent to the trefoil. An argument similar to that in Example 5.22 beginning on p. 128 shows that it is algebraic and nonsingular.

Example 5.27. We mentioned in Example 5.25 that the four-leaved rose in Figure 5.22 has equation

$$(x^2 + y^2)^3 = x^4 - 6x^2y^2 + y^4.$$

Its degree is 6 and its order at the origin is 4; the singularity is ordinary of order four. Any line though the origin not tangent to the curve there intersects the curve in two non-origin points in $\mathbb{C}^2$ symmetric with respect

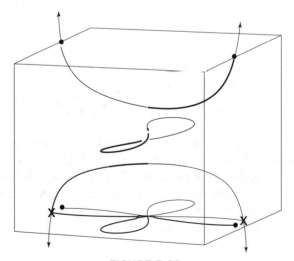

FIGURE 5.23.

5.17. Resolving Several Ordinary Singularities

to the origin. Substituting $y = tx$ into the curve's equation leads to

$$x^2 = \frac{1 - 6t^2 + t^4}{(1 + t^2)^3}.$$

This makes sense: any line through the origin intersects the curve in two complex points having x-components that differ only in sign. Using t as a tag lifts the two points to the t-level. Figure 5.23 shows the space curve along with the original rose in the plane as the shadow at the bottom of the surrounding coordinate box. In this real view, the heavily drawn parts correspond to choosing the negative square root, and the lightly drawn to the positive.

The two dark dots on the space curve project down to the corresponding dots on the plane rose, and the **x**'s show where the space curve trivially projects to the same point on the plane curve. Some lines $y = tx$ appear to intersect the real rose in only the origin because the other two points have nonreal coordinates.

Example 5.28. Sometimes the rotating line method can remove nonordinary singularities. A famous example is the cusp $y^2 = x^3$ which, when lifted this way, has general point (t^2, t^3, t). Choosing the t-axis as domain makes the curve the graph of the smooth vector function (t^2, t^3). The method also works for any curve $y^m = x^n$ where m and n are relatively prime.

5.17 Resolving Several Ordinary Singularities

The rotating line method works for a single ordinary singularity, but what about a curve with two or more of them? The Lissajous figures on p. 101 suggest how easy it is to manufacture curves having millions of nodes. Fortunately, the line slope approach can be made to work for any number of ordinary singularities. We sketch the method here, leaving it to the reader to consult a more detailed treatment such as the one in Chapter 7 of [Fulton].

The rotating line method gives a simple, intuitive way of presenting the lifting idea, sometimes called "blowing up a point." Importantly, the set of all complex lines through a fixed point P covers each point of $\mathbb{C}^2 \setminus \{P\}$ exactly once. But we could equally well let a point (x, y) wander about, visiting each point of $\mathbb{C}^2 \setminus \{P\}$ exactly once. Whenever it encounters a point of the curve, record the point's location along with its complex slope

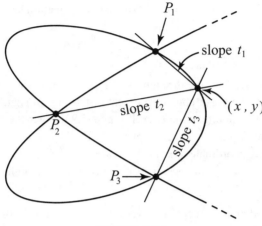

FIGURE 5.24.

with respect to P. This slope, as a tag to form a triple, lifts the point from $\mathbb{C}^2$ to $\mathbb{C}^3$. This wandering point idea directly generalizes to several points.

Let's suppose, therefore, that the plane curve C has N ordinary singularities at $P_1, \cdots, P_N$, and let (x, y) be a point wandering about, visiting each point of $\mathbb{C}^2 \setminus \{P_1, \cdots, P_N\}$ exactly once. When (x, y) encounters a point of C, record its location along with the (usually complex) N slopes t_i with respect to each of the N points P_i. Add all N tags to (x, y) to make the $(N+2)$-tuple $(x, y, t_1, \cdots, t_N)$, which represents a lifting of (x, y) into $(\mathbb{C}^2 \setminus \{P_1, \cdots, P_N\}) \times \mathbb{C}^N$. The end result is that the topological closure in $\mathbb{C}^2 \times \mathbb{C}^N$ of this lifting turns out to be a nonsingular algebraic curve. The part of the lifted curve near a lifting of P_i is the graph of a vector-valued holomorphic function.

Figure 5.24 shows the situation for a curve C having three nodes. For $i = 1, 2, 3$, the complex number t_i is the slope of the line through P_i and (x, y). The point (x, y) on the plane curve gets lifted to the point $(x, y, t_1, t_2, t_3) \in \mathbb{C}^5$. For convenience, we can choose coordinates so that none of the N lines has infinite slope.

5.18 Quadratic Transformations

Quadratic transformations are an especially symmetric type of birational transformation. We don't go into detail, but just outline their main features. They are important because they can transform any singularity of a plane curve into an ordinary singularity, while keeping the transformed curve in the plane. In fact, there's always some sequence of quadratic transfor-

5.18. Quadratic Transformations

mations making all singularities ordinary. After all singularities have been transformed into ordinary ones, the method of the last section can be used to desingularize the curve, although the resulting nonsingular curve may be in a much higher-dimension space and may also have a higher degree. As a final step, however, it is always possible to project this nonsingular curve in a high-dimensional space into $\mathbb{C}^3$ while keeping it nonsingular. (See, for example, Chapter VI, Sec. 4.4 of [Walker].)

Quadratic transformations are defined very simply on the projective plane, so let (x, y, z) be homogeneous coordinates defining a point in $\mathbb{P}^2(\mathbb{C})$.

Definition 5.14. The *standard quadratic transformation of* (x, y, z) is

$$(yz, xz, xy).$$

It maps all but three points of $\mathbb{P}^2(\mathbb{C})$ to $\mathbb{P}^2(\mathbb{C})$, the exceptions being what might be called "the three origins" — the three points where the coordinate axes $x = 0$, $y = 0$ and $z = 0$ intersect pairwise. These are points having two zero entries, so their computed image is $(0, 0, 0)$, which is not a point of $\mathbb{P}^2(\mathbb{C})$ but rather the empty set there. At points of $\mathbb{P}^2(\mathbb{C})$ off the three coordinate axes, the transformation is its own inverse since then xyz isn't zero and applying the transformation twice to the projective point (x, y, z) gives the same projective point:

$$(x, y, z) \to (yz, xz, xy) \to (xzxy, yzxy, yzxz) = xyz(x, y, z),$$

this last determining the same projective point as (x, y, z).

If a curve C is defined in $\mathbb{P}^2(\mathbb{C})$ by a homogeneous polynomial $h(x, y, z)$, we can apply the quadratic transformation to the argument (x, y, z) to get $h(yz, xz, xy)$. In keeping with the name "quadratic," this transformation doubles the degree of h. To carry out the quadratic transformation on an affine curve, first homogenize $p(x, y)$ to $h(x, y, z)$, find $h(yz, xz, xy)$ and, if desired, dehomogenize at $z = 1$ to get its affine image.

The following example illustrates the computation, and shows a nonordinary singularity becoming ordinary.

Example 5.29. The curve defined by $x^4 - y^2 - y^4$ has lowest-degree part y^2, so the origin is a nonordinary point, the line tangent there being double. This is reflected in the left picture in Figure 5.25. The homogenization of the defining polynomial is $h(x, y, z) = x^4 - y^2 z^2 - y^4$, and applying the standard quadratic transformation to it gives

$$(yz)^4 - (xz)^2(xy)^2 - (xz)^4 = z^2(y^4 z^2 - x^4 y^2 - x^4 z^2).$$

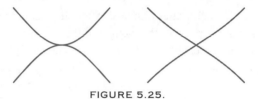

FIGURE 5.25.

Setting $z = 1$ gives
$$y^4 - x^4 y^2 - x^4.$$

Its real part appears in the right picture of Figure 5.25. The singularity looks ordinary, but *is* it? The lowest-degree part $y^4 - x^4$ factors as

$$(y - x)(y + x)(y - ix)(y + ix),$$

revealing that in addition to the tangent lines of slope ± 1, there are two others of slope $\pm i$. Since the four slopes are different, the singularity is indeed ordinary of order 4.

For more on quadratic transformations, see [Coolidge], p. 196–212, [Fulton], Chapter. 7, Sec. 4, or [Walker], p. 74–86, as well as p. 137 there. See [B-K] for a clear and extensive discussion of desingularizing curves.

We end this chapter with a picture of a real curve having a generous allotment of cusps and nodes. At the beginning of Chapter 1, we mentioned that curves obtained by rolling one along another can generate algebraic curves. Hypocycloids, for example, are generated by a point on a circle as it rolls without slipping along the inside of a larger circle. If the circles' radii are rationally related, the hypocycloid is an algebraic curve and can be parametrized by

$$\{x = m\cos(nt) + n\cos(mt), \ y = m\sin(nt) - n\sin(mt)\}.$$

These and other roulettes played an important historical role because by using them, the ancient Greeks were able to include in their models of heavenly movements the vexing, mysterious retrograde motions of the "wanderers" — the planets. As a result of their wrong-headed geocentric approach to the problem, they became veritable experts in a range of roulettes.

In Figure 5.26, $a = 25$ and $b = 19$. This and others like it can be shown to be algebraic using the same kind of argument in Example 1.2 on p. 10: convert x and y^2 to polynomials in $T = \cos t$ and use the resultant of these polynomials to eliminate T, obtaining a polynomial $p(x, y)$ of large degree.

5.18. Quadratic Transformations

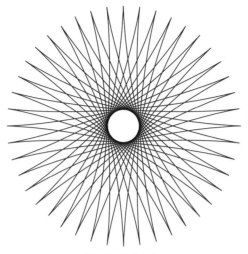

FIGURE 5.26.

CHAPTER **6**

THE BIG THREE: C, K, S

There are three central players in our subject. Although to the unsuspecting they may appear quite different, the unreasonable truth is that they're one and the same, each in different clothing. Without all the proper definitions just yet, they are

- C, an irreducible curve in $\mathbb{P}^2(\mathbb{C})$;
- K, a field of transcendence degree 1 over $\mathbb{C}$;
- S, a compact Riemann surface which, for the moment, can be thought of as a nonsingular curve in $\mathbb{P}^3(\mathbb{C})$.

Each of these three has a notion of equivalence, and there are equivalences from any one to any other.

Uniting the apparently dissimilar is nothing new to science. Uncovering unsuspected relationships is a hallmark of scientific progress. Examples:

- Descartes discovered the connection between Euclidean geometry and algebra, two huge branches of mathematics that for many centuries had led mostly separate lives. His coordinate system allowed us to translate between geometry and much of algebra. This relation eventually expanded to algebraic geometry, of which algebraic curves is a part.

- Before Newton, there was on the one hand "terrestrial physics" and on the other, "celestial physics." His force laws and Universal Law of Gravitation united them into one physics.

- Darwin uncovered the kinship between various forms of life, and in modern times this kinship has been extended to show DNA overlap between virtually any two forms of life — a broad and enlightening unity.

- Einstein's $E = mc^2$ established his famous link between mass and energy, previously thought to be separate. The same is true of Minkowski's insight that Einstein's Special Theory implied a single entity, space-time;

space and time were considered unrelated before that. Even mass-energy and space-time became quantitatively linked in Einstein's General Theory.

In algebraic curves, not only do C, K and S share an essential unity. In a natural way there are mappings between them that fall into a diagram. Each of C, K and S has a natural notion of equivalence that divides them into equivalence classes $\overline{C}$, $\overline{K}$ and $\overline{S}$. Figure 6.1 shows the two versions:

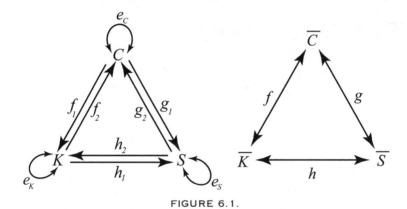

FIGURE 6.1.

In the next sections we explain Figure 6.1. What we say will vary from a cursory reminder of something we've met earlier in this book, to more detailed comments and examples surrounding new material. Some of the results we meet are deep and have nontrivial proofs. In the interests of space we often just state results, giving references to proofs in other books.

About the organization in explaining the figure: The left diagram describes the more concrete, while the right one describes the more general. For example in the left diagram, C represents the totality of individual algebraic curves in $\mathbb{P}^2(\mathbb{C})$, and a map such as f_1 associates to each curve a particular field. The left diagram therefore describes a universe of concrete examples. Similarly for K and S. A map such as e_C denotes birational equivalence, and leads to the diagram on the right by dividing the collection of curves into classes $\overline{C}$ of mutually birationally equivalent curves. Similarly, e_K gathers fields into mutually isomorphic classes $\overline{K}$, while e_S partitions all compact Riemann surfaces into subcollections $\overline{S}$ of conformally equivalent Riemann surfaces. (See Definition 6.5 on p. 146.) The maps f, g and h are now naturally and uniquely determined, each being 1:1 and onto.

We discuss the vertices of the diagrams first, then edges giving relations

6.1 FUNCTION FIELDS

Definition 6.1. A field K has *transcendence degree* 1 over $\mathbb{C}$ if it is an algebraic extension of $\mathbb{C}(t)$, where t is an indeterminate. Therefore K has the form $\mathbb{C}(t, \alpha_1, \ldots, \alpha_r)$, where each α_i satisfies a polynomial equation with coefficients in $\mathbb{C}(t)$.

In the context of algebraic curves, we have:

Definition 6.2. A *function field* K is any field of transcendence degree 1 over $\mathbb{C}$.

For any field $\mathbb{C}(t, \alpha_1, \ldots, \alpha_r)$ in Definition 6.2, the Theorem of the Primitive Element tells us that there exists a single α algebraic over $\mathbb{C}(t)$ so that $\mathbb{C}(t, \alpha_1, \ldots, \alpha_r)$ is isomorphic to $\mathbb{C}(t, \alpha)$. For a proof see, for example, pp. 126–7 of [van der Waerden, vol. I].

Definition 6.3. An equivalence $e_K : K_1 \longleftrightarrow K_2$ is a field isomorphism between function fields K_1 and K_2 that is the identity on $\mathbb{C}$. We call this a $\mathbb{C}$-isomorphism.

Notation. We denote by $\overline{K}$ the set of all equivalence classes of function fields under e_K.

Comment 6.1. We will see later that a genus g can be attached to any function field. It turns out that there is only one equivalence class having $g = 0$, while the equivalence classes of fields of $g = 1$ can be parametrized by two real variables, and those function fields of a particular $g > 1$ can be parametrized by $6(g - 1)$ real variables.

Example 6.1. We met two function fields in Example 5.19 on p. 122: the field $\mathbb{C}(t)$ of a line, and the field $\mathbb{C}(x, y)$ of the alpha curve, with y dependent on the indeterminate x via the alpha curve's irreducible polynomial equation. In Example 5.20 on p. 124, we met the function field $\mathbb{C}(x, y)$, where this time y depends on x through a circle's defining irreducible polynomial. We showed in these examples that the fields of the alpha curve and of the circle are both isomorphic to $\mathbb{C}(t)$. The line has genus $g = 0$ and because there is just one equivalence class having $g = 0$, these isomorphisms show that the alpha curve and the circle, whose equations define curves in $\mathbb{P}^2(\mathbb{C})$, also have genus 0.

6.2 COMPACT RIEMANN SURFACES

We have seen in Chapter 4 that a nonsingular projective curve is a compact, orientable 2-manifold of genus g — that is, a topological sphere with g handles. But it has more structure than just a topological manifold. By the Implicit Function Theorem (Theorem 4.3 on p. 81), about each of its points the nonsingular curve is locally the graph of some complex analytic function, and this additional structure allows us to give the manifold a locally complex-analytic structure, yielding a compact Riemann surface. Here is the definition.

Definition 6.4. A *compact Riemann surface* S is a compact 2-manifold together with a collection $\{U_i, \phi_i\}$ satisfying
- The U_i are countably many open sets covering S.
- Each ϕ_i is a homeomorphism from U_i to an open set $\phi_i(U_i) \subseteq \mathbb{C}$.
- For any i, j for which $U_i \cap U_j \neq \emptyset$, $y = \phi_j(\phi_i^{-1}(x))$ biholomorphically maps $\phi_i(U_i \cap U_j)$ onto $\phi_j(U_i \cap U_j)$.

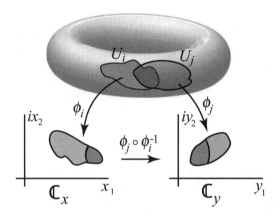

FIGURE 6.2.

We can think of the ϕ_i^{-1}, ϕ_j^{-1} as defining local complex coordinates in each U_i, U_j, and that whenever $U_i \cap U_j \neq \emptyset$, their local coordinates are biholomorphically related. Note that by the definition of a 2-manifold, a Riemann surface is topologically connected. (See (Definition 4.7 on p. 84.) An argument like that in section 4.6 shows that a Riemann surface is orientable.

Definition 6.5. A 1:1 onto mapping $e_S : S \longleftrightarrow S'$ is a *conformal equivalence* between Riemann surfaces S and S' if and only if S and S' have

6.2. COMPACT RIEMANN SURFACES

structure-defining coverings $\{U_i\}$ and $\{U_i'\} = \{e_S(U_i)\}$ so that e_S is biholomorphic between the $\phi_i(U_i)$ and $\phi_i'(U_i')$. Conformally equivalent Riemann surfaces are also called *biholomorphically equivalent*.

Notation. We denote by $\overline{S}$ the set of all equivalence classes of compact Riemann surfaces under the equivalence e_S.

Comment 6.2. It is easy to check that any complex projective nonsingular curve inherits the structure of compact Riemann surface through its embedding in $\mathbb{P}^2(\mathbb{C})$.

We now look at a few specific constructions of compact Riemann surfaces.

Example 6.2. We can give an ordinary sphere the structure of compact Riemann surface. Choose as a model the sphere of diameter 1 centered at the origin of $\mathbb{R}^3$, and cover the sphere by two open sets: U_1 is the sphere minus the north pole and U_2 is the sphere minus the south pole. In Figure 6.3, a copy of $\mathbb{C}$ is tangent to the sphere at the south pole, and another copy

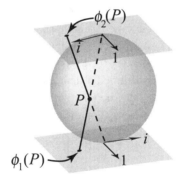

FIGURE 6.3.

of $\mathbb{C}$ slides along a line of longitude to the north pole where it's still tangent but upside down. Let ϕ_1 be the projection from the north pole to the plane tangent to the south pole. Let ϕ_2 be the projection from the south pole to the plane tangent to the north pole. These projections associate any point P on the sphere corresponding to $x = re^{i\theta}$ on the lower plane, with $y = \frac{1}{r}e^{-i\theta}$ on the upper plane. The map connecting the two local coordinate systems on the sphere is $y = \frac{1}{x}$, which is biholomorphic on $\phi_1(U_1 \cap U_2)$.

Example 6.3. We've said that there are infinitely many conformally distinct compact Riemann surfaces of genus 1, their equivalence classes being parametrized by two real variables. Although we show here just one way of giving a torus the structure of Riemann surface, our construction imparts the flavor of others. To begin, think of the torus as the topological product of two circles and cut it along two intersecting circles so the surface can be unwrapped and laid out on a plane as a rectangle. Figure 6.4 shows the cut torus starting to unwrap, with the oriented cut edges labeled a and b. The

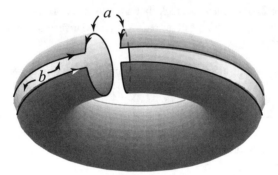

FIGURE 6.4.

torus can be represented as $\mathbb{R}^2$ modulo the lattice generated by the rectangle vertices. Figure 6.5 depicts the rectangle as the square having diagonal vertices $(0, 0)$ and $(1, 1)$, so the lattice generated by its vertices consists of integer pairs (m, n). The plane is tiled this way by unit squares, and any two points in the plane are identified if their coordinates differ by integers. In any one square, opposite sides are identified, and the identified sides correspond to one of the two circles cut on the torus.

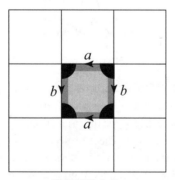

FIGURE 6.5.

This cutting and the identifications naturally define a splitting up of the torus in Figure 6.4 — or equally well, of the closed unit square — into four

6.2. COMPACT RIEMANN SURFACES

disjoint sets. They are: 1 open face, 2 open intervals (edges) and 1 point (the 4 identified vertices). These four sets can be covered by four open sets in the plane: an open square, two open rectangles, and one open disk. This can be seen in Figure 6.5, keeping in mind the identifications. If the picture is appropriately laid over the complex plane, then the identity map $y = x$ serves as the biholomorphic map between any two overlapping open sets, thus turning the torus into a Riemann surface.

Example 6.4. We extend Example 6.3, showing how to construct infinitely many conformally distinct compact Riemann surfaces of genus 1. Although a torus is the product of two circles, we've said nothing about the relative sizes of the circles. They can differ greatly — think of a garden hose with its ends screwed together. The torus would then be represented as a tiling of the plane by congruent strip-like rectangles. In fact, rectangles could be replaced by congruent parallelograms corresponding to, say, horizontally shearing the picture in Figure 6.5. In any case, the same type of covering may be used, the picture being drawn on $\mathbb{C}$ and the identity map $y = x$ defining biholomorphic maps. Note that "biholomorphic" implies *conformality* — the map is angle-preserving. So although any parallelogram defines a Riemann surface, the resulting Riemann surfaces may not be conformally equivalent. For example, vertically compress Figure 6.5 to half its height. Under this compression, most angles change, some decreasing in size, others increasing, so the two associated Riemann surfaces are not conformally equivalent.

Example 6.5. A compact orientable topological manifold of any genus can be made into a Riemann surface. We've already done this for $g = 0$ and $g = 1$. To proceed inductively, we increase by one the genus of any compact orientable topological manifold by cutting a hole in both the manifold of genus g and a torus, and then gluing together the cut edges to obtain a manifold of genus $g + 1$. Figure 6.6 depicts doing this to get a genus 2 manifold from two tori.

The top row shows two tori, each with a hole cut out. Notice the labeling showing how opposite edges are to be identified, with arrows agreeing in direction. All four vertices of each square are identified to one point, the point where the two circles on the original torus met. In the second row, each hole's edge is straightened, and now five points are identified to one. In the bottom picture, the straightened edges have been glued together, the whole figure forming a regular polygon of $8 = 4g$ sides. Identifying sides so that arrow directions agree produces a manifold of genus 2. To make this

150   6. THE BIG THREE: C, K, S

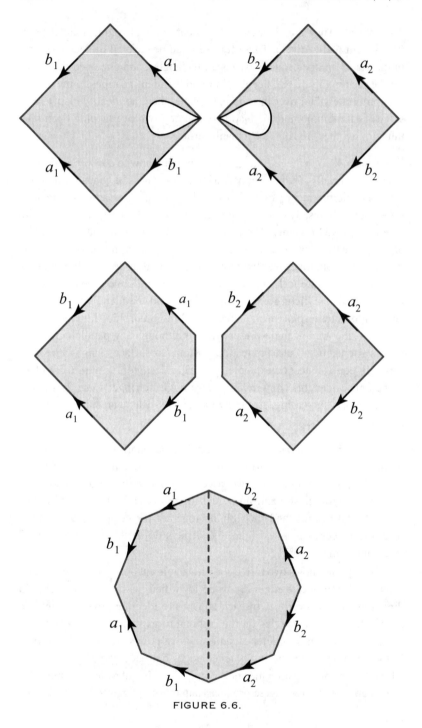

FIGURE 6.6.

6.2. COMPACT RIEMANN SURFACES

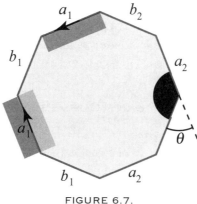

FIGURE 6.7.

into a Riemann surface, we need to define the maps ϕ_i. Figure 6.7 indicates the idea.

On the open octagonal interior, use the identity map $y = x$ with an underlying copy of $\mathbb{C}$. For a typical side, the two-tone open rectangle attached to the lower a_1 can similarly be assigned the identity map. Then trim off the darker half and sew it back on the upper a_1 as indicated, respecting arrows. Since the two sides are identified, the result is still a connected open set on the manifold. As for the vertices, the angular opening of the dark pie-shaped area is $\pi - \theta = \pi - \frac{2\pi}{4g} = \frac{\pi(2g-1)}{2g}$. There are $4g$ of these for a total angle of $2\pi(2g-1)$. Since the $4g$ vertices are identified to a single point, that one point has an angle of $2\pi(2g-1)$ around it. The map $y = x^{2g-1}$ obligingly multiplies any angle by $2g - 1$, as required. In Figure 6.7, use $y = x^{2g-1} = x^3$ translated to each vertex and applied to the pie-shaped area there. Notice that we used the map $y = x^{2g-1}$ on the rectangle model of the torus, because in that case $y = x^{2g-1}$ reduces to $y = x$. We did the same for the sphere, since $g = 0$ means $y = x^{-1}$.

To keep things in perspective, the above example assumes the $4g$-gon is regular so it constructs only *one* specific Riemann surface for each $g > 0$. As in turning a rectangle into a compact Riemann surface, applying a linear transformation that changes the shape of the $4g$-gon changes angles within it, and therefore leads to different Riemann surfaces of genus g. Since there is just one conformal class of compact Riemann surface of genus 0, the associated parameter space of conformal classes consists of just one point. On the other hand, the conformal classes of compact Riemann surfaces of genus 1 are parametrized by a parameter space of real dimension 2. For a sneak preview of what parameter spaces can look like, see Figure 6.11 on

p. 162. It turns out that for $g > 1$, the parameter space has real dimension $6g - 6$. (See, for example, (4.12) of [Clemens] or Chapter. XI, Lemma 4.9 of [Miranda].) Example 6.5 indicates how to construct a Riemann surface corresponding to just one point in each of these parameter spaces.

6.3 PROJECTIVE PLANE CURVES

We have defined *irreducible projective curve* $C \subset \mathbb{P}^2(\mathbb{C})$ in Definition 3.2 on p. 46, taking C to be the set of complex 1-subspaces of $\mathbb{C}^3$ in the zero set of the homogenization $h(x, y, z)$ of $p(x, y)$. Alternatively, C may be looked at as the topological closure in $\mathbb{P}^2(\mathbb{C})$ of a curve defined in $\mathbb{C}^2 \subset \mathbb{P}^2(\mathbb{C})$ by an irreducible polynomial $p(x, y)$. To simplify things, we will assume coordinates have been chosen so that C is not the line at infinity.

Equivalence e_C in Figure 6.1 is birational equivalence. We defined the function field of an affine curve in Definition 5.10 on p. 119 and birational equivalence in Definition 5.12 on p. 120. These definitions have projective analogs:

Definition 6.6. Let C be the zero set of an irreducible homogeneous polynomial $h = h(x, y, z)$. Then (h) consisting of all $\mathbb{C}[x, y, z]$-multiples of h is a prime ideal in $\mathbb{C}[x, y, z]$. Let $\mathfrak{D}$ be the integral domain $\mathbb{C}[x, y, z]/(h)$. Any element of $\mathfrak{D}$ can be represented by some homogeneous polynomial in $\mathbb{C}[x, y, z]$, and every nonzero element of $\mathfrak{D}$ is represented by some homogeneous polynomial relatively prime to h. The *function field of an irreducible curve* $C \subset \mathbb{P}^2(\mathbb{C})$ is the field of quotients g/h, where g and h are homogeneous of the same degree and h is nonzero in $\mathfrak{D}$.

Definitions 5.10 and 6.6 in fact give the same function field, in the sense of this basic result:

Theorem 6.1. The function field of an irreducible curve $C \subset \mathbb{P}^2(\mathbb{C})$ is isomorphic to the function field of any curve in $\mathbb{C}^2 \subset \mathbb{P}^2(\mathbb{C})$ whose topological closure is C.

For a proof, see [Kunz], Theorem 4.4, pp. 34–5.

From Definition 5.12 on p. 120, curves C_1 and C_2 are birationally equivalent if and only if their function fields K_1 and K_2 are $\mathbb{C}$-isomorphic. We write $e_C : C_1 \longleftrightarrow C_2$.

The diagrams in Figure 6.1 are suggestive. We used the left side of each diagram in defining birational equivalence of curves, but we can equally well use the right side, looking instead at conformally equivalent Riemann surfaces instead of isomorphic function fields. Here's a definition equivalent to Definition 5.12.

Definition 6.7. Curves C_1 and C_2 are birationally equivalent if and only if their Riemann surfaces S_1 and S_2 are conformally equivalent. We again write $e_C : C_1 \longleftrightarrow C_2$

As we've noted before, a birational map between two curves is not always 1:1 on points of the curves. For example, we saw in Chapter 5 that in desingularizing a node, the branches through the node separate under the birational map into branches having distinct centers. However, Theorem 5.7 on p. 127 tells us something important: any birational map from one projective curve to another maps in a 1:1 onto manner the places of the first curve to those of the other. We discuss this further in the next section.

Notation. We denote by $\overline{C}$ the set of all birational equivalence classes of curves under the equivalence e_C defined in either Definition 5.12 or Definition 6.7.

A basic fact about $\overline{C}$: A birational transformation applied to a curve keeps the curve's image in $\overline{C}$ fixed, its function field isomorphism class in $\overline{K}$ fixed, and its Riemann surface conformal class in $\overline{S}$ fixed. At every stage in desingularizing a curve, all images in the right diagram of Figure 6.1 stay put. This applies as well to projecting a nonsingular curve in a high dimensional space to a nonsingular model in $\mathbb{P}^3(\mathbb{C})$. In particular, at each stage of desingularizing, the genus never changes. For positive genus, the point in the space parametrizing Riemann surfaces of a given genus never changes.

6.4 f_1, f_2, f : Curves and Function Fields

Definition 6.6 assigns to any curve $C \subset \mathbb{P}^2(\mathbb{C})$ a field of rational functions on C. Let f_1 denote such an assignment. We can think of f_1 as mapping any element of $\overline{C}$ to an element of $\overline{K}$, defining in this way f in the right diagram of Figure 6.1. This map f sends $\overline{C}$ onto $\overline{K}$, because the Theorem of the Primitive Element tells us that any function field κ is isomorphic to one of the form $\mathbb{C}(x, y)$, where x is an indeterminate and y satisfies a dependence equation $p(x, y) = 0$. The topological closure in $\mathbb{P}^2(\mathbb{C})$ of the curve $C(p) \subset \mathbb{C}^2$ then maps to the class of κ in $\overline{K}$. Also, $f : C/e_C \to K/e_K$ is 1:1 by the definition of birational equivalence: two curves are in the same birational equivalence class exactly when they have the same function field, up to isomorphism. Therefore f has an inverse f^{-1}; in the right diagram we write f with a two-sided arrow.

As for f_2, let $K = \mathbb{C}(t, \alpha)$, where α satisfies an irreducible polynomial equation with coefficients in $\mathbb{C}(t)$. Suppose we choose complex values for both t and α so that the polynomial equation for α remains satisfied. Substituting these values into each element of K defines a $\mathbb{C}$-homomorphism of K into $\mathbb{C}$. Conversely, any $\mathbb{C}$-homomorphism of K into $\mathbb{C}$ corresponds to choosing such values for t and α. The collection of all the assigned ordered pairs defines an affine curve in $\mathbb{C}^2$, and taking its topological closure in $\mathbb{P}^2(\mathbb{C})$ defines an irreducible curve $C \subset \mathbb{P}^2(\mathbb{C})$. We say that f_2 maps the function field K to the curve C. The function fields isomorphic to K map to curves birationally equivalent to C. This is expressed in the right diagram by the upward arrow of f.

The maps f_1, f_2 and f actually do more than link curves with function fields. They can link even individual parts of curves with parts of function fields. The "micro-parts" of a curve are *places*, and the parts of the function field are *valuation subrings* of the field. Here's the idea. First, in defining intersection multiplicity in Theorem 3.2 on p. 56 (Chapter 3), we used the notion of order of a polynomial at a place: we took a place represented by a parametrization $\{x = t^r, y = \text{power series in } t\}$, substituted it into a polynomial and considered the order of t in the result. For any particular such parametrization we can apply this process to elements of a projective curve's function field K and in this way, the place defines an *order function* on the field. This can be regarded as a group homomorphism from the multiplicative group of $K \setminus \{0\}$ to the additive group of integers, $\mathbb{Z}$. This homomorphism can be extended to a *discrete valuation* v by defining the order of $0 \in K$ to be ∞. The following easily-verified properties can be used to define a general valuation. For $a, b \in K$:

- $v(ab) = v(a) + v(b)$,
- $v(a + b) \geq \min(v(a), v(b))$,
- $v(0) = \infty$.

Associated to any valuation is a *valuation ring* V consisting of all $a \in K$ for which $v(a) \geq 0$. Intuitively, V consists of the functions in K which don't assume the value ∞ at the place corresponding to the valuation ring. It is easily checked that for any element $a \in K$, we have either $a \in V$, $\frac{1}{a} \in V$, or both. This property can be taken as the definition of a *valuation ring of K*.

Any place of a projective curve with function field K defines a valuation ring. This association is 1:1 and onto. For onto, any valuation ring of K must come from some place of C. It turns out that any valuation ring uniquely defines an analytic parametrization which in appropriate coordinates can be taken to have center $(0, 0)$:

$$\{x = t^r, \; y = f(t) = \text{a power series in } t\}.$$

This parametrization can be constructed inductively; see the proof of Theorem 10.3 of [Walker], pp. 158–9. The well-defined nature of the construction also shows that the association is 1:1.

A birational map from one curve to another is defined by an isomorphism between their function fields, and such an isomorphism carries valuation rings to valuation rings, so also places to places. Since there's a 1:1 onto correspondence between places of C and valuation rings in K, we see that a birational correspondence e_C between two projective curves induces a 1:1 onto correspondence between the places of one curve and those of the other — the content of Theorem 5.7 on p. 127.

6.5 g_1, g_2, g: COMPACT RIEMANN SURFACES AND CURVES

The map g_1 attaches to any curve $C \subset \mathbb{P}^2(\mathbb{C})$ a compact Riemann surface. Furthermore, as a consequence of the Riemann-Roch theorem, which we meet later in this chapter, we have:

Theorem 6.2. *Every compact Riemann surface is conformally equivalent to a nonsingular curve C in $\mathbb{P}^3(\mathbb{C})$.*

One example of a g_1 is sending a curve in $\mathbb{P}^2(\mathbb{C})$ to a desingularization of it in $\mathbb{P}^3(\mathbb{C})$; this g_1 is then a function on the set of all curves in $\mathbb{P}^2(\mathbb{C})$ because, as we saw in Chapter 5, any such curve does have a desingularization. g_1 is a birational map from an irreducible curve in $\mathbb{P}^2(\mathbb{C})$ to a compact Riemann surface.

Theorem 6.3. *For any particular g_1, there exists a finite set of points of the irreducible curve off which g_1 is not only well-defined but biholomorphic (that is, conformal). This finite set includes the set of singularities of the curve.*

This is a consequence of Theorem 9.3 on p. 169 in [Fischer].

As for g_2, Theorem 6.2 together with a suitable projection into $\mathbb{P}^2(\mathbb{C})$ insures that g_2 is defined on S in the left diagram of Figure 6.1.

g is 1:1 and onto between $\overline{C}$ and $\overline{S}$, because from Definition 6.7, if two curves aren't birationally equivalent, then their Riemann surfaces aren't conformally equivalent. Theorem 6.2 furthermore tells us that g is onto $\overline{S}$.

This last has quantitative implications that can be surprising. For instance, Example 5.16 on p. 114 says there are plane curves of any genus. We've also mentioned that for $g > 1$, the conformally distinct compact Riemann surfaces of genus g are parametrized by $6(g - 1)$ real variables, so this number increases linearly with $g - 1$. To show just how fast this number can grow, think back to the Fermat curve $x^n + y^n = 1$ mentioned in Examples 1.15 and 1.17 on pp. 26 and 27. For $n = 1000$, in $\mathbb{R}^2$ the curve is smooth yet bends in four places so rapidly that it looks like a square. In $\mathbb{P}^2(\mathbb{C})$ the surface has $\frac{999 \cdot 998}{2} = 498,501$ holes in it, and requires an n-tuple consisting of $6 \cdot 498,500 = 2,991,000$ real numbers to specify the conformal equivalence class of the Riemann surface — a real n-tuple with nearly 3 million components. One can push the numbers even further: every compact Riemann surface of genus g is the Riemann surface of some curve in $\mathbb{P}^2(\mathbb{C})$ of degree $2(g - 1)$ that has $2(g^2 - 4g + 3)$ nodes; see [Miranda], p. 70. Therefore if we're handed a plane curve C and told only that it has the same genus as the above Fermat curve, we can deduce that C's defining polynomial could have degree as high as 997,000, and the curve might have as many as 1,988,010,024,006 (nearly two trillion) nodes.

6.6 h_1, h_2, h: Function Fields and Compact Riemann Surfaces

We may define h_1 as the composition $g_1 \circ f_2$. Since both f_2 and g_1 are 1:1 and onto, so is the composition, and h_2 is the inverse of h_1. Then h maps between equivalence classes of function fields and conformal classes of compact Riemann surfaces.

6.7 Genus

The concept of genus arises in all three of C, K, and S and their quotient spaces. In each, it is a fundamental invariant. Here are some informal definitions and facts.

- The genus g of a compact Riemann surface is its genus as a topological manifold — that is, as a sphere with g handles. It is a conformal invariant.

- Topologically, an irreducible projective algebraic curve is an oriented, compact manifold $\mathfrak{M}$ of genus g, but possibly modified by identifying finitely many of its points to finitely many points. The genus of the curve is the genus of $\mathfrak{M}$ and is a birational invariant.
- The genus of a function field may be defined as the genus of any associated Riemann surface or projective curve. It is invariant under field isomorphism.

6.8 Genus 0

Knowing the genus in any of the three contexts of C, K, or S tells us quite a bit. The following theorem summarizes many of the basic facts about curves of genus zero. The mutual equivalences can be established through results in [Kunz], Chapter 14, [Miranda], Chapter VII § 1 and [Walker], Chapter V, § 7.

Theorem 6.4. For an irreducible curve $C \subset \mathbb{P}^2(\mathbb{C})$, these five statements all equivalent:
- C has genus 0.
- C is birationally equivalent to $\mathbb{P}^1(\mathbb{C})$.
- The function field of C is isomorphic to $\mathbb{C}(t)$, t an indeterminate.
- The Riemann surface of C is the Riemann sphere.
- C has a rational parametrization.

If a genus zero curve is in addition nonsingular, then substituting $g = 0$ into the genus formula

$$g = \frac{(n-1)(n-2)}{2}$$

shows that the curve's degree n is 1 or 2. We therefore have

Theorem 6.5. If a nonsingular curve $C \subset \mathbb{P}^2(\mathbb{C})$ has genus 0, then C is either a line or a nondegenerate conic.

In Examples 5.19 (p. 122) and 5.20 (p. 124) of rational parametrizations, each curve has genus 0. By Theorem 6.4, so do all irreducible cusp curves $y^m = x^n$ discussed in section 5.9 starting on p. 115, since $y^m = x^n$ has the rational parametrization $\{x = t^m, \ y = t^n\}$. In fact, we can directly show that they're all birationally equivalent to the projective line — that the fields $\mathbb{C}(x, y)$ and $\mathbb{C}(t)$ are isomorphic. For this, express t rationally in terms of

x and y: since m and n are relatively prime, their gcd is 1, so we can write $1 = ma + nb$ for some integers a and b. Therefore $t = t^1 = t^{ma+tb} = t^{ma}t^{nb} = (t^m)^a(t^n)^b = x^a y^b$.

If the projective curve $C(p)$ defined by $p(x, y)$ is not topologically a sphere with finitely many points identified to finitely many points, then there exists no parametrization

$$\{x = r_1(t), \, y = r_2(t)\}$$

of $C(p)$ in which r_1 and r_2 are rational.

If we specialize the right diagram in Figure 6.1 to the case of $g = 0$, then Theorem 6.4 tells us that the diagram is trivial — there's just one element in each of $\overline{C}$, $\overline{K}$, and $\overline{S}$. The situation for $g = 1$ is far different. Its story makes up one of the major chapters of mathematics whose roots can be traced back as early as Diophantus of Alexandria (3rd century A.D.) and is still unfinished. In the next section, we present a diagram that encapsulates some highlights of the story.

6.9 Genus One

This section centers around a nonsingular genus one analog of the diagrams in Figure 6.1.

Definition 6.8. A projective nonsingular curve of genus 1 is called an *elliptic curve*.

Elliptic curves are the simplest possible curves after lines and conics. Their study has flowered into a whole field in which geometry, algebra and complex analysis combine and illuminate one another. Though we don't touch upon it, more recently number theory has also entered the picture, greatly affecting the whole landscape of this "queen of science." For example, elliptic curves played a central role in the proof of Fermat's Last Theorem by Andrew Wiles (assisted by Richard Taylor). Because so much specific information is available in genus 1, the diagrams expand in a natural way to include four main objects as shown in Figure 6.8.

E **and** $\overline{E}$ **:**

The letter E stands for an elliptic curve. We've already met an example in Chapter 5: the curve defined by $y^2 = (x + 1)x(x - 1)$ in Example 5.11 on p. 103. Its sketch in the real plane appears in Figure 5.3, and the curve is featured again in Figure 5.4, (a), (b). Figure 6.9 depicts this curve in the real projective plane. Notice how the branches meet at the end of the y-axis.

6.9. GENUS ONE

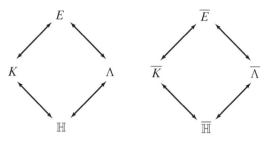

FIGURE 6.8.

In the context of elliptic curves, this point at infinity is usually denoted by O. (We'll see why when we give E a group structure on p. 165.)

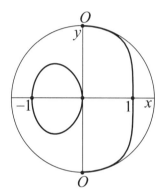

FIGURE 6.9.

We note these things about this complex curve:
- It is nonsingular.
- It is defined by a cubic of the form $y^2 = x^3 + ax + b$.
- The three roots of $0 = x^3 + ax + b$ are distinct.
- The point at infinity is "rational," having projective coordinates $(0, 0, 1)$.

Comment 6.3.

- Since an elliptic curve is projective, nonsingular and has genus 1, the genus formula becomes $\frac{(n-1)(n-2)}{2} = 1$, which simplifies to $n(n-3) = 0$. Therefore any elliptic curve is defined by some polynomial of degree 3.
- By appropriately changing affine coordinates, any polynomial equation $y^2 = \{\text{a cubic in } x\}$ can be put into the form $y^2 = x^3 + ax + b$. A simple criterion for nonsingularity turns out to be

$$\left(\frac{a}{3}\right)^3 + \left(\frac{b}{2}\right)^2 \neq 0.$$

Curves such as the top, middle and bottom ones in the right column of Figure 1.3 on p. 8 have equations in this standard form.

- By appropriately changing projective coordinates by applying a nonsingular linear transformation of $\mathbb{C}^3$, it turns out that a general cubic in x and y for an elliptic curve can be put into the form $y^2 = x(x-1)(x-\lambda)$, where $\lambda \in \mathbb{C} \setminus \{0, 1\}$. $\lambda \neq 0, 1$ insures that the curve in nonsingular.
- All six curves shown in Figure 1.1 on p. 5 are elliptic curves because coefficients of the general cubic p were randomly chosen. Since the singularity conditions $p_x = p_y = 0$ are satisfied with probability 0, the curves are all nonsingular. None of these curves are standard-form ones such as the top, middle and bottom sketches in the right column of Figure 1.3, but each can be made so by an appropriate change of coordinates in $\mathbb{C}^3$.
- E, as a subset of $\mathbb{P}^2(\mathbb{C})$, inherits the structure of a Riemann surface. Any elliptic curve is its own Riemann surface.

Definition 6.9. Two elliptic curves are *isomorphic* if the homogeneous polynomial defining one can be transformed into that of the other by a linear change in (x, y, z)-coordinates in $\mathbb{C}^3$.

Here is an alternative form of Definition 6.9:

Definition 6.10. Two elliptic curves are *isomorphic* if the curves, considered as homogeneous sets in $\mathbb{C}^3$, are connected through a nonsingular linear transformation of $\mathbb{C}^3$.

Either definition partitions the set of all elliptic curves into equivalence classes $\overline{E}$.

$\Lambda, \overline{\Lambda}, \mathbb{H}, \overline{\mathbb{H}}$

Λ denotes a lattice in $\mathbb{C}$ consisting of all integer linear combinations $m\omega_1 + n\omega_2$ of two $\mathbb{R}$-linearly independent elements ω_1, ω_2 of $\mathbb{C}$. The idea is that ω_1 and ω_2 form adjacent sides of a parallelogram determining a "fundamental domain" of the lattice. Identifying opposite sides of the parallelogram defines a torus whose Riemann surface structure is determined by Λ. (See Examples 6.3 and 6.4 on pp. 147 and 149.) For any elliptic curve E, there always are complex numbers ω_1, ω_2 generating a lattice Λ whose Riemann structure is conformally the same as the Riemann surface E. The lattice accomplishing this is not unique because uniformly changing scale or rotating the lattice preserves angles and thus has no effect on the structure of the induced Riemann surface. That is, multiplying Λ by any fixed

6.9. GENUS ONE

nonzero complex number c yields the same Riemann surface. Lattices Λ and Λ' are often called *homothetic* if $\Lambda' = c\Lambda$ for some such c, but we will call them simply *similar*.

Definition 6.11. Lattices Λ and Λ' are *similar* if $\Lambda' = c\Lambda$ for some $c \in \mathbb{C} \setminus \{0\}$.

Even for a given Λ, a basis $\{\omega_1, \omega_2\}$ is never unique. Any nonsingular matrix $\begin{pmatrix} a & b \\ c & d \end{pmatrix}$ whose entries are integers maps integer combinations $m\omega_1 + n\omega_2$ to integer combinations, and therefore Λ into (but not necessarily onto) Λ. Multiplying the original area by the determinant of the matrix gives the area of the image parallelogram. Therefore if in addition the matrix is unimodular (has determinant ± 1), then the parallelogram area remains unchanged. Any lattice parallelogram is a basis parallelogram if and only if it has minimum nonzero area, so a unimodular matrix maps a Λ-base to a Λ-base and therefore maps $\Lambda \to \Lambda$ in a 1:1 onto way. In any Λ, there are therefore infinitely many dissimilar basis parallelograms. Figure 6.10 shows two of them.

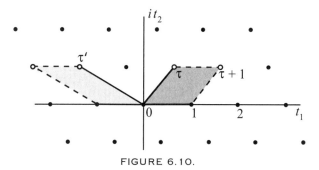

FIGURE 6.10.

By dividing any lattice basis $\{\omega_1, \omega_2\}$ by ω_1, we can always assume the basis has the form $\{1, \frac{\omega_2}{\omega_1}\}$, and by replacing $\frac{\omega_2}{\omega_1}$ by $-\frac{\omega_2}{\omega_1}$ if necessary, we can further assume $\frac{\omega_2}{\omega_1}$ lies in the upper half-plane $\mathbb{H} = \{c \in \mathbb{C} | \Im(c) > 0\}$. In this way, a single complex number $\tau = \frac{\omega_2}{\omega_1} \in \mathbb{H}$ determines a lattice basis. Since there are many such numbers τ, it is natural to ask how they are related. Suppose that $\tau = \frac{\omega_2}{\omega_1}$ and that $\tau' = \frac{\omega'_2}{\omega'_1}$, where $\frac{\omega'_2}{\omega'_1} = \frac{a\omega_1 + b\omega_2}{c\omega_1 + d\omega_2}$. Dividing numerator and denominator by ω_1 yields the linear fractional transformation $\tau' = \frac{a+b\tau}{c+d\tau}$. Looking directly at this fraction, we see that the coefficients in numerator and denominator assemble themselves into the unimodular matrix $\begin{pmatrix} a & b \\ c & d \end{pmatrix}$; from the fraction we can write the matrix and from the matrix we can write the fraction. If we restrict ourselves to the modular group Γ of unimodular matrices having determinant $+1$, then all images of τ remain in the upper half plane $\mathbb{H}$.

We know what a typical element of Γ looks like from an algebraic viewpoint — it's a 2×2 unimodular matrix of determinant $+1$. But Γ is also generated by easily visualizable geometric transformations: for any $z \in \mathbb{C}$, both translation $T : z \to z + 1$ and the map $S : z \to -\frac{1}{z}$ are linear fractional and map $\mathbb{H}$ to itself. Γ itself consists of all possible monomials in T and S under composition, so the orbit of any point P consists of rows of integral translates and multiple applications of S. There are natural regions $F \subset \mathbb{H}$ consisting of one representative from each orbit under Γ, which we call the *fundamental domains* of $\mathbb{H}$ under the action of Γ. Γ maps the fundamental domains to each other, and in this way Γ divides $\mathbb{H}$ into equivalence classes $\mathbb{H}/\Gamma$. These equivalence classes are shown in Figure 6.11.

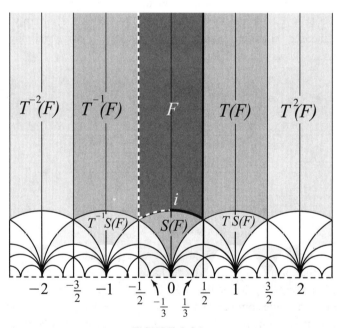

FIGURE 6.11.

The darkest region is called the *canonical fundamental domain*, with successively lighter shades representing image regions under monomials in T and S of successively higher degrees. It turns out that in a basis $\{1, \tau\}$, τ can always be chosen to lie in the canonical fundamental domain. In the diagram of Figure 6.8, $\mathbb{H}/\Gamma$ is denoted by $\overline{\mathbb{H}}$.

6.9. GENUS ONE

K and $\overline{K}$:
The meanings of K and $\overline{K}$ are like those in Figure 6.1 — any element is a function field or isomorphism class of function fields of an underlying curve defined by an irreducible polynomial. However, with elliptic curves we can say more.

Relations Between Objects in Figure 6.8

In the following bulleted items, we state without proof some of the highlights of a cluster of facts about elliptic curves.

• As noted earlier, any elliptic curve E can be defined by a polynomial equation of the form
$$y^2 = 4x^3 + ax + b$$
for some $a, b \in \mathbb{R}$. (See, for example, [B-K], Chapter II, §7.3, Theorem 11.) We always assume $\left(\frac{a}{3}\right)^3 + \left(\frac{b}{2}\right)^2 \neq 0$ to insure that E is nonsingular. In that case $y^2 - (4x^3 + ax + b)$ is irreducible and defines a quotient field, as in Definition 5.10 on p. 119. But for an elliptic curve, we can explicitly write down a doubly periodic meromorphic function on $\mathbb{C}$ — a Weierstrass $\wp$-function — such that the function field of $E = \mathbb{C}(y^2 - 4x^3 - ax - b)$ is $\mathbb{C}(\wp(t), \wp'(t))$. A function f on $\mathbb{C}$ is called *doubly periodic* with *periods* ω_1 and ω_2 provided ω_1 and ω_2 are $\mathbb{R}$-linearly independent complex numbers and $f(x) = f(x + n_1\omega_1 + n_2\omega_2)$ for all integers n_1, n_2.

Here's the explicit function:
$$\wp(t) = \frac{1}{t^2} + \sum_{k=1}^{\infty} c_{k+1} t^{2k}, \tag{6.1}$$

where
$$c_2 = -\frac{a}{20}, \quad c_3 = -\frac{b}{28},$$
and for indices $k > 3$,
$$c_k = \frac{3}{(2k+1)(k-3)} \sum_{i=2}^{k-2} c_i c_{k-1}.$$

(See [A-S], p. 635.) From the exponents of t, we see that the function is even: $\wp(-t) = \wp(t)$ for all $t \in \mathbb{C}$.

• The derivative $\wp'(t)$ exists, is odd, and is doubly periodic with the same periods as $\wp$. Since t is an indeterminate, $\mathbb{C}(\wp(t))$ has transcendence degree 1 over $\mathbb{C}$. Since E is a curve, the transcendence degree of

$\mathbb{C}(\wp(t), \wp'(t))$ over $\mathbb{C}$ is 1, so $\wp'(t)$ is algebraic over $\mathbb{C}(\wp(t))$. Writing $\wp$ for $\wp(t)$, $\wp$ satisfies the differential equation

$$\wp'^2 = 4\wp^3 + a\wp + b.$$

(See [Cartan], Chapter V, §2.5 or [Kunz], Chapter 10.) Therefore $\wp'$ is algebraic over $\mathbb{C}(\wp(t))$.

- $\wp(t)$ and its derivative $\wp'(t)$ parametrize E. That is,

$$\{x = \wp(t), y = \wp'(t)\}$$

maps a fundamental parallelogram in $\mathbb{C}$ to the torus E by identifying sides of a parallelogram-shaped region like those in Figure 6.10. (See [Cartan], Chapter V, §2.5, Proposition 5.2 or [Kunz], Chapter 10.) As t runs through the points of a fundamental region, the points of E are covered once. We will see in the next section how this is analogous to parametrizing a complex circle with singly-periodic trigonometric functions:

$$\{x = \cos(t), y = \sin(t)\}.$$

- The leading term of the expansion of $\wp(t)$ in (6.1) shows it has a double pole at the origin, and its parallelogram-shaped fundamental domain of definition fits in with its being doubly periodic. In fact, its doubly periodic array of poles defines a lattice Λ, which is therefore uniquely determined by the coefficients a and b. The dependence of $\wp$ and Λ on a and b can be emphasized by writing $\wp(t|a,b)$ and $\Lambda(a,b)$.

- In the above, we obtained the function field K starting from E. We can also get $\wp(t)$, and therefore the function field of E by starting with a lattice Λ. In the following series in $t \in \mathbb{C}$, λ runs over all nonzero points of Λ:

$$\wp(t) = \frac{1}{t^2} + \sum_{\lambda \neq 0}\left(\frac{1}{(t-\lambda)^2} - \frac{1}{\lambda^2}\right).$$

(See [Hartshorne], p. 327.)

- Above, we derived the lattice Λ from the coefficients a and b in $y^2 = 4x^3 + ax + b$. We can go the other way, too, obtaining a and b, and therefore the curve's equation, from the nonzero elements λ in a given lattice Λ. Here are the formulas:

$$a = -60 \sum_{\lambda \neq 0} \lambda^{-4} \quad \text{and} \quad b = -140 \sum_{\lambda \neq 0} \lambda^{-6}.$$

6.9. Genus One

For a proof, see [H-C], Chapter II. For background reading and perspective, see [Hartshorne], Chapter IV, §4.

• Vector addition in $\mathbb{C}$ defines an abelian group structure not only on $\mathbb{C}$, but on the torus $\mathbb{C}/\Lambda(\tau)$, and this can be transferred to E via the parametrization $\{x = \wp(t), y = \wp'(t)\}$. As a group, E is an *abelian variety*, though the term "abelian curve" would be appropriate, too. This group structure can also be defined in a purely geometric way. Let P and Q be any two points of the projective curve $E \subset \mathbb{P}^2(\mathbb{C})$. Since E has degree three, Bézout's theorem tells us that the line through P and Q intersects E in one other point, say with coordinates (x, y). (If $P = Q$, take the "limiting line" — the tangent line — through P.) Then define $P + Q$ to be $(-x, y)$, which is the reflection about the x-axis of this intersection point. Figure 6.12 shows the idea in the real setting. Notice something unusual when we choose Q

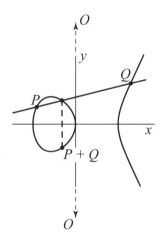

FIGURE 6.12.

to be the point at infinity (the point O in Figure 6.9 on p. 159). The line through P and Q is vertical, so the intersection point (x, y) is just the reflection of P about the x-axis. Therefore the reflection of this reflection is again P — that is, the point at infinity serves as the zero element of the group, explaining why we denoted point at infinity by O in Figure 6.9. It's easy to check that the additive inverse $-P$ is then the reflection of P about the x-axis.

This simple reflection method of defining a group law on the elliptic curve forces the curve's zero element to be the point at infinity, but actually, *any* point of the elliptic curve can be selected to be the identity element O. The construction is essentially the same, with the line L though the two points O and (x, y) replacing the vertical line as the "reflector." That is, the sum $P + Q$ is the remaining (third) intersection of L with the cubic. However, no matter what the choice of O, we still need to establish the associative law to insure that we have a group. This is the most involved step in a proof. For this, and a discussion of group structures on cubic curves, see [Kunz], Chapter 10 or [Reid], Chapter 1, §2.

6.10 AN ANALOGY

There is a suggestive analogy between the parametrization $\{\wp(t), \wp'(t)\}$ of an elliptic curve E and the parametrization $\{\cos t, \cos' t\}$ of a complex circle C. We choose language and notation to highlight the analogy.

• Associated to C is the quadratic $q = y^2 + x^2 - 1$, a standard form such that any nonsingular quadratic curve is birationally equivalent to $C(q)$. C is a complex circle of unit radius.

• Associated with the lattice $\Lambda' = \{2\pi n \mid n \in \mathbb{Z}\}$ in the real axis of $\mathbb{C}$ is what we may call a canonical fundamental domain of Λ', depicted in Figure 6.13. Any parallel translate of this region by an element of Λ' is a

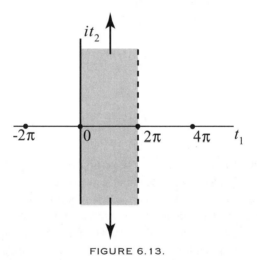

FIGURE 6.13.

fundamental domain.

• $\mathbb{C}/\Lambda'$ is a Riemann surface of genus 0 (a sphere), and since all Riemann surfaces of genus 0 are conformally equivalent, $\mathbb{C}/\Lambda'$ is conformally equivalent to the Riemann surface C.

6.10. AN ANALOGY

- $\cos t$ defined on $\mathbb{C}$ is singly periodic, having the canonical fundamental domain as a period strip. $\cos t$ is even; its derivative $\sin t$ is odd and singly periodic with the same fundamental domain as $\cos t$.
- The circle $C = C(q) \subset \mathbb{C}^2$ is parametrized by

$$\{x = \cos t, \, y = \cos' t\}.$$

As t runs through points of a fundamental domain, points of C as well as its points at infinity are covered once.

- The function field of C is $\mathbb{C}(\cos t, \cos' t)$. Its transcendence degree over $\mathbb{C}$ is 1, since $\sin^2 t = -\cos^2 t + 1$ implies that $\cos' t = -\sin t$ is algebraic over $\mathbb{C}(\cos t)$. This is also expressed as the differential equation

$$(\cos' t)^2 = -(\cos t)^2 + 1.$$

Of course, $\mathbb{C}(\cos t)$ is isomorphic to $\mathbb{C}(t)$.

Let's look more closely at a parametrization of an elliptic curve versus that of a circle.

Elliptic curve. Each of Figures 6.14, 6.15, and 6.16 shows four canonical loops on a torus, with real loops drawn solid and imaginary loops being dashed. *Mathematica* greatly helps in exploring specific elliptic curves. Instead of the standard form $y^2 = x^3 + ax + b$, *Mathematica* uses the closely related classical Weierstrass normal form $y^2 = 4x^3 - g_2 x - g_3$. The command **WeierstrassInvariants[{0.5, 0.5 I}]** tells us that the two real values $g_2 \approx 189.073$ and $g_3 = 0$ produce half-periods of $\frac{1}{2}$ and $\frac{i}{2}$, making the full period parallelogram the unit square in $\mathbb{C}$ with $\tau = i$, as depicted in Figure 6.14. Four canonical line segments in this square respectively map under the parametrization

$$\{x = \wp(t), \, y = \wp'(t)\}$$

into the real branch, real loop, imaginary branch, and imaginary loop of the four loops of $C(y^2 - 4x^3 + g_2 x + g_3)$ seen in (x_1, y_1, iy_2)-space. Figure 6.14 shows four directed unit line segments, each one unit long. Their ends are identified, meaning they form four oriented loops. Figure 6.15 shows these oriented loops as they lie on the part of $C(p)$ in (x_1, y_1, iy_2)-space. Figure 6.16 shows them as they lie on the torus. Note how the two branches meet at a common point at infinity.

Complex Circle. Figures 6.14, 6.15, and 6.16 have analogs for the complex circle; they are Figures 6.17 and 6.18. The part of the complex unit

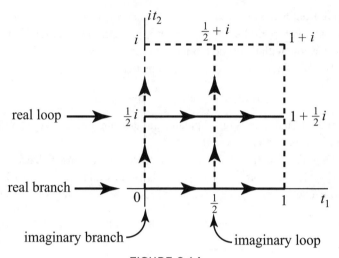

FIGURE 6.14.

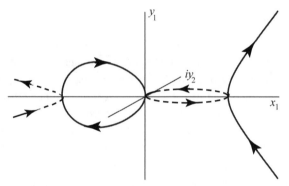

FIGURE 6.15.

circle $C(q)$ in (x_1, y_1, iy_2)-space consists of a real circle in the (x_1, y_1)-plane together with the hyperbola $x_1^2 - y_2^2 = 1$ in the (x_1, iy_2)-plane. Under the parametrization $\{x = \cos t, y = \cos' t\}$, the directed loop from $t_1 = 0$ to $t_1 = 2\pi$ in Figure 6.17's fundamental strip maps into the real circle with clockwise orientation. Going up the strip's line $t_1 = 0$ to the point at infinity and then down the line $t_1 = \pi$ represents a continuous loop and corresponds to traversing the right branch in the indicated direction to infinity, meeting the other branch there and returning along it. Continuing along the left branch, we head towards another point at infinity. On these branches, the parametrization

$$\{x = \cos t, y = \cos' t\}$$

6.10. AN ANALOGY

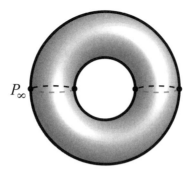

FIGURE 6.16.

becomes

$$\{x = \cos it_2, y = -\sin it_2\}.$$

Using angles $\theta = \pm it_2$ in Euler's formula $e^{i\theta} = \cos\theta + i\sin\theta$ shows that $\cos it_2 = \cosh t_2$ and $\sin it_2 = i\sinh t_2$. Also, addition formulas lead to $\cos(\pi + it_2) = -\cosh t_2$ and $\sin(\pi + it_2) = -i\sinh t_2$. So just as circular functions parametrize the circle in the picture, hyperbolic functions parametrize the hyperbolic branches. The complex circle has genus 0, and

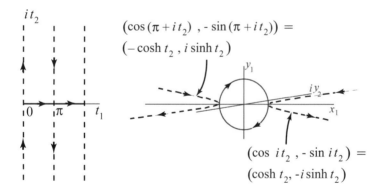

FIGURE 6.17.

we can think of the real circle as being the sphere's equator and the two branches as being antipodal lines of longitude (semicircles) on the sphere. If we could visualize in four dimensions, we'd see a hyperbola coming off each antipodal point-pair of the real circle, each hyperbola corresponding to two lines of longitude forming a great circle on the sphere. Figure 6.18

shows the solid equatorial loop corresponding to the real circle, and the dotted loop corresponding to the hyperbola branches meeting in two points on $\mathbb{P}^2(\mathbb{C})$'s line at infinity in which, by Bézout's theorem, the complex circle is guaranteed to intersect. Of course, the hyperbolic loop was topologically contorted to make it lie on a sphere in $\mathbb{R}^3$. The figure also depicts two other branches coming off two other antipodal points of the real circle, forming another loop and once again passing through the two points at infinity.

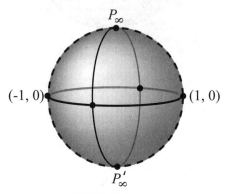

FIGURE 6.18.

6.11 Equipotentials and Streamlines

A nonconstant polynomial $p(x, y)$ defines not only an algebraic curve $C(p)$ in $\mathbb{C}^2$. It also defines a decomposition or "fibration" of $\mathbb{C}^2$ consisting of disjoint algebraic curves $p(x, y) = c, c \in \mathbb{C}$, thus splitting up $\mathbb{C}^2$ into a union of curves of complex dimension 1. We can do an analogous thing at the real level, splitting up a nonsingular complex curve into a disjoint union of real curves, plus possibly finitely-many points. The real curves can be interpreted as "equipotentials," or dually, "streamlines." (See [Needham] for illuminating reading in the case of genus 0. This book contains an abundance of good pictures; his last three chapters are especially relevant.)

> **Notation.** Our focus will from now on be on real one-dimensional curves, so henceforth we use notation familiar in this context: $z = x + iy$ and $w = u + iv$. This requires fewer subscripts.

Definition 6.12. A twice continuously-differentiable function $u(x, y)$: $\mathbb{R}^2 \to \mathbb{R}$ is called *harmonic* on $\mathbb{R}^2$ provided $u_{xx} + u_{yy} = 0$ at each point of $\mathbb{R}^2$. A second function $v(x, y) : \mathbb{R}^2 \to \mathbb{R}$ harmonic on $\mathbb{R}^2$ is called

6.11. EQUIPOTENTIALS AND STREAMLINES

conjugate to $u(x, y)$ provided the two functions are the real and imaginary parts of a holomorphic function $w : \mathbb{C} \to \mathbb{C}$. We write such a function w as $w(z) = u(x, y) + iv(x, y)$. The level curves of u and v intersect orthogonally at any point of $\mathbb{R}^2$ where the complex derivative w' is nonzero, and the collection of these curves is called an *orthogonal net*.

Example 6.6. We can split the holomorphic map $w = z$ into parts as $u(x, y) + iv(x, y) = x + iy$. The conjugate harmonic functions $u(x, y)$ and $v(x, y)$ are x and y; the level curves of x are vertical lines and those of y are horizontal lines. Think of the plane as covered by a thin sheet of incompressible fluid, and let the function x represent a "velocity potential." This means that the gradient $\nabla x = (1, 0)$ evaluated at each point $P \in \mathbb{C}$ defines a vector field representing the velocity of the fluid at P. In this example, the vectors are constant — all are unit vectors pointing rightward. Accordingly, the fluid flows horizontally at a steady rate from left to right along streamlines $y = $ a constant. At any P in a streamline, the vector at P is tangent to the streamline there. In all this, the roles of u and v can be reversed, with $\nabla y = (0, 1)$ defining a vector field everywhere orthogonal to $\nabla x = (1, 0)$, the fluid then flowing upward along lines $x = $ a constant. The two sets of level curves form an orthogonal net.

Example 6.7. Let $w = z^3$. The real and imaginary parts of $w = u + iv$ are $u(x, y) = x^3 - 3xy^2$ and $v(x, y) = 3x^2y - y^3$. Figure 6.19 shows the level-curves of $u(x, y)$ on the left and those of $v(x, y)$ in the middle. Their union on the right depicts the level curves of $u(x, y)$ orthogonally intersecting those of $v(x, y)$ at all points except the origin, where $w' = 0$.

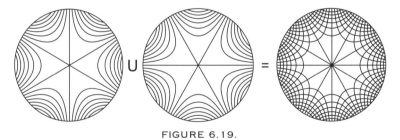

FIGURE 6.19.

If we choose $u(x, y) = x^3 - 3xy^2$ to be the velocity potential, then the left picture depicts equipotential lines where u is constant, and the gradient $\nabla u(x, y) = 3(x^2 - y^2, -2xy)$ defines a vector field representing the velocity of the fluid at each point (x, y). The middle picture depicts the resulting streamlines. In the pie-shaped region in the first quadrant of the middle picture, the fluid flows clockwise, and the sense alternates as we move from one pie-shaped region to the next. The roles of u and v can be reversed. In that case the left picture represents streamlines and the center picture, curves along which v is constant.

These ideas form a basic part of complex analysis of one variable, where the emphasis is on functions defined on $\mathbb{C}$ or the Riemann sphere. The notion of function field of a nonsingular projective curve as a compact Riemann surface means we can do complex analysis on the surface. Because it is constructed from patches of $\mathbb{C}$, it makes sense to talk about the real and imaginary parts of a holomorphic or meromorphic function, and to say that they are harmonic. We can therefore carry over to compact Riemann surfaces the notions of equipotential lines and streamlines orthogonal to them.

Let's explore these ideas on a compact Riemann surface of genus 1. In

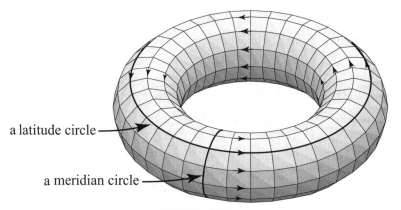

FIGURE 6.20.

the figure, the small circles on the torus are the meridians and are orthogonal to the latitudes, and the orthogonal net may be regarded as the image of rectangular coordinate lines in $\mathbb{C}$ under the mapping from a rectangle in $\mathbb{C}$ to the torus, with edges identified in the usual way. In $w = u + iv = z$ on the complex plane, the contour curves along which the harmonic function $u(x, y)$ is constant, are the torus meridians. Conjugate to u is the harmonic function v whose streamlines are the circles of latitude on the torus.

We can think of fluid velocity as proportional to force, so that moving a point Q against the fluid flow requires work. We can calculate the amount of work done in moving Q along a path γ on the torus by integrating the vector component of fluid velocity along the path. Suppose P is a fixed reference point on the torus. In the figure, if we move along a latitude

6.11. EQUIPOTENTIALS AND STREAMLINES

against the depicted current to another point Q on the torus, positive work is done; moving with the current corresponds to performing negative work. The roles of u and v can be interchanged, with v being the potential function. In that case, we can write iv instead of v, and moving along a meridian circle corresponds to doing pure imaginary work. In general, moving from P to any other point then requires complex work.

By moving, say, no more than one turn in either the latitude or meridian direction, the complex numbers representing work fill out a region R of $\mathbb{C}$, but not all of it since the amount of work done is bounded. However, the restriction of moving less than 360° is artificial. If c_1 represents the real work done in going once around a latitude circle from P to P and c_2 is the imaginary work done from P to P by going around a meridian circle, then the complex work done in going from P to Q along any path is $w(Q) + \lambda_1 c_1 + \lambda_2 c_2$ for integers $\lambda_i \in \mathbb{Z}$. The set $\{\lambda_1 c_1 + \lambda_2 c_2 \mid \lambda_i \in \mathbb{Z}\}$ is a lattice Λ in $\mathbb{C}$ and $R \times \Lambda$ covers $\mathbb{C}$. Thus w is multiple-valued on the torus, its values differing by elements of Λ.

This torus example generalizes to any compact Riemann surface S of genus g. Suppose we have penciled on S $2g$ loops so that if we were to cut along them, we'd obtain the $4g$-gon featured in Example 6.6 starting on p. 149. But instead of cutting, turn each loop into a rubber band, constrained always to lie on S. The surface and rubber band are considered to be frictionless. There will be g cross points $P_1, \cdots, P_g$, each being where some two rubber bands intersect. Pull all g points P_i to a common point and identify them to that one point. This point corresponds to the identified vertices of the $4g$-gon. In this pulling, the rubber bands move, too. Because the rubber bands are frictionless and constrained to stay on S, at the end of this process the assemblage of rubber bands will assume a position of minimum energy. That is enough to guarantee that each is an equipotential line $u_i =$ constant or $v_i =$ constant of g complex potentials $w_1, \cdots, w_g$.

It can be shown that the set of $2g$ potentials $\{u_1, \cdots, u_g, v_1, \cdots, v_g\}$ is linearly independent over $\mathbb{R}$ up to an additive constant, in the sense that if

$$\alpha_1 u_1 + \cdots + \alpha_g u_g + \alpha_{g+1} v_1 + \cdots + \alpha_{2g} v_g \equiv \text{constant},$$

then all $2g$ of the α_i must be 0. (See, for example [Springer], p. 28 or [Klein], p. 39.) We can then construct a flow of the incompressible fluid so that we expend nonzero energy to go around one circle but not around any of the others. Figure 6.21 is an example of one such flow on a surface of genus 3. It requires positive work to traverse the leftmost streamline circle clockwise, but an algebraic total of no energy to go around either of the

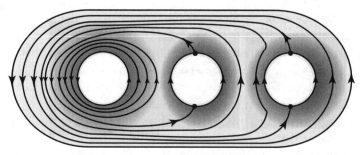

FIGURE 6.21.

other two circles. Since the remaining 6 − 3 circles intersect the streamlines orthogonally, there is no component of force along them, so no work is required to go around them, either. Notice the two solid dots on each of the two right circles; these are branch points where there is a fork in the streamline and the fluid splits up or rejoins. Increasing the genus by adding more holes and additional circles in the picture would add two more such branch points per hole. The total number of branch points is easily seen to be $2g - 2$.

Such elementary flows can be linearly combined to create more general flows on compact Riemann surfaces. For example, on the torus one building block flow follows meridian circles and another follows latitude circles. So a pure meridian flow that is ten times faster than a pure latitude flow results in streamlines where a particle spirals in the meridian direction ten times as it moves once around in the latitude direction. At the end of the trip the particle returns to its starting position because one and ten are commensurable — rational multiples of each other. If the speed in one direction were, say, π times that in the other, the spiral would never close up on itself, instead spiraling around from the infinite past to the infinite future, never self-intersecting.

6.12 Differentials Generate Vector Fields

The differential $f(z)dz$ generates two orthogonal vector fields as follows: write $f(z) = u + iv$ and $dz = dx + idy$. Then

$$f(z)dz = (u + iv)(dx + idy) = (udx - vdy) + i(vdx + udy) =$$
$$(u, -v) \cdot (dx, dy) + i(v, u) \cdot (dx, dy).$$

Therefore the differential $f(z)dz$ generates the two fields

$$\big(u(x, y), -v(x, y)\big) \quad \text{and} \quad \big(v(x, y), u(x, y)\big).$$

6.13. A Major Difference

They are orthogonal since their dot product $(u, -v) \cdot (v, u)$ is zero.

Example 6.8. To obtain the two vector fields arising from the differential $z^3 dz$, write $z^3 = u + iv = (x^3 - 3xy^2) + i(3x^2y - y^3)$. The vector fields are then
$$(u, -v) = (x^3 - 3xy^2, y^3 - 3x^2y)$$
and
$$(v, u) = (3x^2y - y^3, x^3 - 3xy^2).$$
These are seen in the top pictures of Figure 6.22. The bottom picture is their superposition, depicting orthogonality. In all three pictures, the center of the plot is the origin of $\mathbb{C}$.

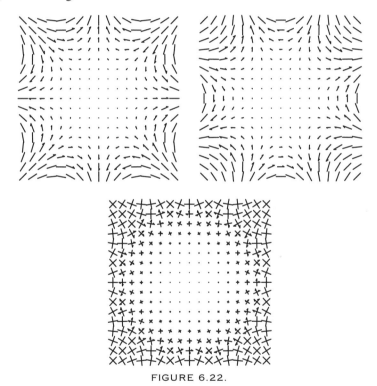

FIGURE 6.22.

6.13 A Major Difference

There are two kinds of differentials: exact and non-exact. For us, an exact differential $f(z)dz$ is one that can be written as $dg(z)$ for some $g(z)$ in the curve's function field. Otherwise we consider the differential to be non-exact.

Some Perspective: A Little about Calculus

In one-variable calculus, the Fundamental Theorem of Calculus has a basic implication: if a function f is continuous on $\mathbb{R}$, then the differential $f(x)dx$ is exact there. That is, any such function has an antiderivative $g(x)$, meaning $g'(x) = f(x)$, or $f(x)dx = dg(x)$. This implication is why first-year calculus students typically don't hear about non-exact differentials.

The theorem also says $\int_a^b f(x)dx = \int_a^b dg(x) = g(b) - g(a)$, and usually the path of integration from a to b is the line segment from a to b. However, the path could just as well wiggle around continuously within the x-axis, moving from a to far beyond to the right or left of a and b, perhaps many times, before finally coming to rest at b. The theorem tells us that all the wiggling amounts to nothing: the value of the integral is still $g(b) - g(a)$. That is, the integral's value is independent of any continuous path from a to b.

The integrals in this chapter can be regarded as calculating work done in a force field, while integrals in a beginning calculus course are often associated with the signed area under a function's graph. The first feels more like physics and the second, more like geometry. Actually, these turn out to be equivalent because a continuous function can be regarded as a force field, and signed area can always be interpreted as signed work. Here's how: a point $(x, f(x))$ on the graph of f determines a vector from $(x, 0)$ to $(x, f(x))$. Rotate this vector about its base counterclockwise 90° so that it lies in the x-axis, where we may now think of it as a force vector. The force vectors form a force field in the x-axis, and a definite integral adds the elements of work done in moving from P to $P + dx$. Using the Fundamental Theorem of Calculus to evaluate the integral amounts to finding a potential energy function such as a height function in a gravitational field (an antiderivative). Work done is then calculated from the net change in energy or height.

In calculus of two variables, however, a force field might induce a whirlpool, and an integration path could encircle it. We can go from point P to the same point P and do no work by not moving, or we could move around the whirlpool once, returning to P and doing an amount of work W. The work done going round and round a single whirlpool along a connected closed path not crossing the whirlpool's center is nW for some $n \in \mathbb{Z}$.

6.13. A MAJOR DIFFERENCE

Example 6.9. On the curve $\mathbb{C}$ the differential $\frac{dz}{z}$ can be written

$$\frac{dz \cdot \overline{z}}{z\overline{z}} = \frac{(dx + i\,dy)(x - iy)}{x^2 + y^2}$$

which separates into the real and imaginary parts

$$\left(\frac{x}{x^2 + y^2}, \frac{y}{x^2 + y^2}\right) \cdot (dx, dy), \left(\frac{-y}{x^2 + y^2}, \frac{x}{x^2 + y^2}\right) \cdot (dx, dy).$$

The vectors in the large parentheses define force fields, the first a source out of which flows an incompressible fluid, and the second, a whirlpool rotating counterclockwise. Figure 6.23 depicts the two mutually orthogonal force fields and a few of their streamlines. The strength of the force field

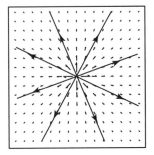 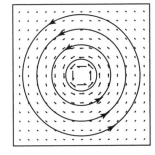

FIGURE 6.23.

and the speed of the fluid decrease as we move away from the field's center of symmetry. On the other hand, the differential $-\frac{dz}{z}$ defines a sink and a clockwise-rotating whirlpool. The antiderivative $\pm \ln(z)$ of either differential $\pm \frac{dz}{z}$ isn't in the function field $\mathbb{C}(z)$ of $\mathbb{C}$ — that is, it isn't a rational function — so relative to $\mathbb{C}(z)$, the differential is not exact. In keeping with this, there are line integrals from a fixed point to other points that do depend on the path and become multiple-valued, a hallmark of a differential being non-exact. In contrast, for $r(z) \in \mathbb{C}(z)$, the exact differential $dr(z)$ has the function $r(z)$ as an antiderivative, and integrating $r(z)$ from a fixed point P to another point Q depends only on Q and not the path — the integral is single-valued.

Example 6.10. For us, differentials on the curve $\mathbb{C}$ all have the form $r(z)dz$, where $r(z)$ is in $\mathbb{C}$'s function field $\mathbb{C}(z)$. Which ones are exact, and which are not? Any differential $\frac{dz}{z-c}$ is non-exact, as is dz times any meromorphic function having one or more simple poles — that is, a function having a (Laurent) series containing a term or terms like $\frac{1}{z-c}$. Integrating around a loop containing a single simple pole produces a nonzero result, the *residue*. It is easy to check that of all terms $(z-c)^n$, only $n=-1$ does this: for all other integer values of n, $(z-c)^n dz$ is exact, and integrating around any closed loop produces zero. (It's assumed that none of these paths cross the point c.)

The characterization of exact versus non-exact differentials in the above example applies to genus 0. What about compact Riemann surfaces of higher genus? If f is any element of the function field K of such a surface, are there non-exact differentials $f dz$ on the surface? The physical arguments given in the section "Equipotentials and Streamlines" starting on p. 170 already go a long way toward answering this question. Consider the torus example in which there are two canonical streamings, one along latitude circles and the other along meridian circles. Moving around a streamline circle requires work, positive or negative, and completing several laps means additional work, so the associated integral is multiple-valued. On the torus, the differential we're integrating could not be exact, for if it were, the integral would be a function — a single-valued antiderivative in K on the Riemann surface. On a Riemann surface of genus g, an analogous argument shows that since there is a basis of g complex differentials on a Riemann surface of genus g, there must be g linearly independent non-exact differentials. This phenomenon is new for us: these differentials are *everywhere finite*, in contrast to the simple poles $\frac{dz}{z-c}$ on the Riemann sphere.

Example 6.11. The curve in $\mathbb{C}^2$ defined by

$$w^2 = (z-a_1)(z-a_2)\cdots(z-a_{2g+1})$$

has genus g provided the a_i are distinct. In the special case when the a_i are the integers from -4 to 4, the equation becomes

$$w^2 = z(z^2-1)(z^2-4)(z^2-9)(z^2-16)$$

and the genus of the curve is four. The intersection of this complex curve with (x, u, iv)-space is shown in Figure 6.24. The four heavily-drawn loops are in the (x, u)-plane and in a natural way depict the holes we see in a closed rubber surface having four holes in a row, and along which we could cut in making the $4g = 16$-gon. The other four loops in the (x, iv)-plane correspond to canonical loop-cuts defining the remaining sides of the 16-gon. It turns out that there are g linearly independent non-exact complex

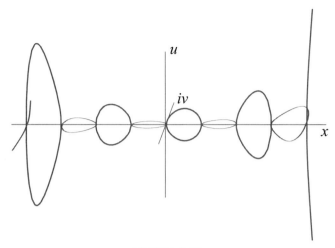

FIGURE 6.24.

differentials on the Riemann surface defined by the curve, and they are

$$\frac{dz}{w}, \frac{zdz}{w}, \frac{z^2 dz}{w}, \cdots, \frac{z^{g-1} dz}{w}.$$

Any non-exact differential on the surface is a linear combination of them. Integrating them represents complex work, and g pairs of analytic loops γ_{i1}, γ_{i2} can be chosen on the surface so that integrating $\frac{z^{i-1} dz}{w}$ around either loop of the pair leads to nonzero work, and to zero work on all the other $2g - 2$ loops.

Example 6.12. When $g = 1$, the curve in Example 6.11 is nonsingular and the part of it in (x, u, iv)-space appears in the picture in Figure 6.15 on p. 168. There is just one non-exact differential, $\frac{dz}{w}$, and Figure 6.20 on p. 172 shows the two conjugate sets of streamlines for it. We have seen on p. 164 that the pair $(\wp, \wp')$ parametrizes the curve, and the function field is isomorphic to $\mathbb{C}(\wp(z), \wp'(z))$.

6.14 Divisors

Keeping track of the number of zeros and poles of a function or differential suggests the concept of *divisor*, which on a compact Riemann surface is simply a formal sum of points-with-multiplicity. For example, take a polynomial whose zeros are points $P_1, \cdots, P_n$ in $\mathbb{C} \cup \{\infty\}$, with P_i having multiplicity m_i. We can write the "divisor of zeros" as

$$\mathfrak{a} = m_1 P_1 + m_2 P_2 + \cdots + m_n P_n.$$

In a specific case like $p(z) = (z-1)^2(z)^6(z+2+i)^5$ defined on $\mathbb{C} \cup \{\infty\}$, its divisor of zeros can be written as

$$2[1] + 6[0] + 5[-2-i].$$

The reciprocal $\frac{1}{p(z)}$ has poles at these points, and a negative multiplicity can denote a pole. Therefore the divisor of poles of $\frac{1}{p(z)}$ is denoted

$$-2[1] - 6[0] - 5[-2-i].$$

Bézout's theorem implies that any polynomial of degree n has n poles, or points at infinity (counted with multiplicity), since the line at infinity has degree 1. Therefore for a polynomial like z^5, the full divisor expressing the placement and multiplicity of both zeros and poles would be $5[0] - 5[\infty]$, while the divisor of $\frac{1}{z^5}$ would be the negative of this, $-5[0] + 5[\infty]$. A general divisor on a compact Riemann surface is written

$$\mathfrak{a} = m_1 P_1 + m_2 P_2 + \cdots + m_n P_n,$$

where the coefficients can be any integers, positive, negative or zero. The divisor's *degree* is $\sum_{i=1}^{n} m_i$, denoted $\deg[\mathfrak{a}]$. Divisors are added just like linear combinations, so if $\mathfrak{a} = 3P_1 - 2P_2$ and $\mathfrak{b} = -P_1 + 2Q_1 + 5Q_2$, then $\mathfrak{a} + \mathfrak{b} = 2P_1 - 2P_2 + 2Q_1 + 5Q_2$. Since any integer can be a coefficient, any divisor has an inverse, obtained by reversing the sign of each P_i; the zero-divisor is defined to be the divisor having all coefficients zero. Therefore all divisors form an abelian group in the expected way. This group even has a partial order: by taking a coefficient m_i to be zero if necessary, we may assume two divisors both have a common form $m_1 P_1 + m_2 P_2 + \cdots + m_n P_n$. Then $\mathfrak{a} \geq \mathfrak{b}$ provided the multiplicity of each point in $\mathfrak{a}$ is equal to or greater than the multiplicity of the same point in $\mathfrak{b}$.

A divisor expressing the placement and multiplicity of all zeros and poles of a meromorphic function f is called a *principal* divisor, and is denoted by (f). If it is $m_1 P_1 + m_2 P_2 + \cdots + m_n P_n$, then m_i is called the *order of f at P_i*. If $(f) \geq \mathfrak{a}$ for a divisor $\mathfrak{a}$, we say that the divisor (f) *cuts out* $\mathfrak{a}$.

Two basic facts:

• Any meromorphic function f on a compact Riemann surface S assumes every value (including infinity) the same number of times. Therefore for any meromorphic function, $\deg[(f)] = 0$.

• On S, the set of all principal divisors forms a subgroup of the group of all divisors. We say two divisors $\mathfrak{a}$ and $\mathfrak{b}$ are *equivalent* if $\mathfrak{a} = (f) + \mathfrak{b}$ for some meromorphic function f on S, and we write $\mathfrak{a} \cong \mathfrak{b}$.

6.14. Divisors

Facts about divisors of meromorphic functions have parallels for divisors of meromorphic differentials. Suppose a differential $\omega = f dz$ has a local Laurent expansion $(c_n z^n + c_{n+1} z^{n+1} + \cdots) dz$ about the point P. If $c_n \neq 0$, we say the order of ω at P is n. The order is nonzero at only finitely many points P_i, so ω defines the divisor $(\omega) = m_1 P_1 + m_2 P_2 + \cdots + m_n P_n$, where the order of ω at P_i is m_i. We call (ω) a *canonical divisor*. If $\omega_1 = f dz$ and $\omega_2 = g dz$, then $\frac{\omega_1}{\omega_2}$ is a meromorphic function, so any two canonical divisors differ by a principal divisor.

Since every principal divisor has degree 0, all canonical divisors on S have the same degree. It turns out that this common degree depends only on the genus of S: $\deg[(\omega)] = 2g - 2$. (See [Fulton], p. 107 or [Kunz], p. 153, for example.) Since every canonical divisor is equivalent to every other one on S, we denote a generic canonical divisor by $\mathfrak{c}$.

Example 6.13. On the Riemann sphere, let $\mathbb{C}$ be a coordinate system about the south pole of the sphere, and consider the differential dz. That's $1 dz$, so the order of the meromorphic function 1 at each point of $\mathbb{C}$ is 0. That leaves one point of the Riemann sphere to consider, ∞. The transformation $z \to \frac{1}{z}$ maps $\mathbb{C}$ to a coordinate system about ∞. Therefore $dz \to d\left(\frac{1}{z}\right) = \frac{-1}{z^2} dz$. This has order -2, so the degree of dz is -2, fitting in with $2g - 2$ since the Riemann sphere has genus 0. This simple differential is representative of all meromorphic differentials as far as degree is concerned. That is, $\deg[(f)] = 0$, so $\deg[(f dz)] = -2$ on the sphere.

Example 6.14. Suppose $g = 1$. The latitude and meridian circles in Figure 6.20 on p. 172 can be taken to depict two sets of streamlines. When viewed in the square with opposite sides identified, one set of streamlines is made up of horizontal gridlines, the other, vertical gridlines. An incompressible fluid flowing along either set has constant nonzero speed, so the differential dz has no zeros or poles, meaning that the order at every point is zero. Therefore in this case $\deg[(dz)] = 0$, which is in keeping with our formula $\deg[(\omega)] = 2g - 2$ when $g = 1$.

Example 6.15. For $g > 1$, Figure 6.21 on p. 174 suggests the basic idea. As noted there, each time we add a hole to increase the genus, the everywhere finite flows must branch at two additional points, resulting in $2g - 2$ branch points. We can think of the incompressible fluid as having net velocity zero at the instant the streamline symmetrically splits apart, and this corresponds to a zero of an everywhere finite differential. The two zeros per hole fits in with $\deg[(\omega)] = 2g - 2$.

6.15 The Riemann-Roch theorem

Compact Riemann surfaces have varying abilities to "hold" meromorphic functions. One way of measuring the capacity is to fix a divisor $\mathfrak{a}$ and ask how many functions f on the surface have principal divisor greater than $-\mathfrak{a}$. For example, on the Riemann sphere let P be the origin, let n be a fixed positive integer, and suppose $\mathfrak{a} = nP$. Then $1, \frac{1}{z}, \cdots, \frac{1}{z^n}$ are $n+1$ linearly independent functions all cutting out $-\mathfrak{a}$. This already generates an $(n+1)$-dimensional vector space over $\mathbb{C}$ since any linear combination also cuts out $-\mathfrak{a}$. For example, suppose $n = 4$. The order of $\frac{1}{z^2} + \frac{1}{z^3}$ is -3, the minimum of -2 and -3, so the sum still cuts out $-\mathfrak{a}$. What about extending the sequence of functions $1, \frac{1}{z}, \cdots, \frac{1}{z^n}$? That is, might either z or $\frac{1}{z^{n+1}}$ also cut out $-\mathfrak{a}$? Not z, because its order at the origin is 1 and its order at infinity is -1. There is no point of $\mathfrak{a}$ at infinity, so the order of $-\mathfrak{a}$ there is 0 and $-1 < 0$. Also not $\frac{1}{z^{n+1}}$, because its order at 0 is $-(n+1)$, which is less than the order of $-\mathfrak{a}$ there.

In this special case, the dimension of the vector space of meromorphic functions cutting out $-\mathfrak{a}$ is $\deg[\mathfrak{a}] + 1$, so $\deg[\mathfrak{a}] + 1$ acts like a credit limit beyond which you may not charge your credit card. It's not hard to generalize from this special $\mathfrak{a}$ to any $\mathfrak{a}$: for $\mathfrak{a}$ on the Riemann sphere,

$$L(-\mathfrak{a}) = \deg[\mathfrak{a}] + 1,$$

where $L(-\mathfrak{a})$ denotes the complex dimension of the vector space of all meromorphic functions cutting out $-\mathfrak{a}$.

The above formula is for $g = 0$. What about a Riemann surface of genus 1? Take a simple case such as a single point P for $\mathfrak{a}$. Then $\deg[\mathfrak{a}] = 1$, so the above formula, if true, would predict that the dimension $L(-\mathfrak{a})$ is $\deg[\mathfrak{a}] + 1 = 2$. The constant functions on the torus form a one-dimensional vector space, so the formula works if we can find a meromorphic function on the torus having a single zero at P and no other zeros, as this would perfectly fulfill the restriction imposed by $\mathfrak{a}$. The catch is,

There is no such function!

There *are* meromorphic functions on a Riemann surface of genus 1 that assume every value twice: the Weierstrass $\wp$-function is an example, and so is its derivative $\wp'$. Since these two functions generate the function field of the Riemann surface, it might be guessed that every function assumes every value at least twice. This is indeed so — there's no meromorphic function that assumes even one value exactly once. This phenomenon arises when we move from a genus zero curve to a genus one curve, so any generalization

6.15. THE RIEMANN-ROCH THEOREM

of our formula must take into account the genus of the underlying Riemann surface. The full answer to this state of affairs is this:

Theorem 6.6. (Riemann-Roch theorem) With notation as above, on a compact Riemann surface of genus g,

$$L(-\mathfrak{a}) = \deg[\mathfrak{a}] + L(\mathfrak{a} - \mathfrak{c}) - g + 1.$$

Compared with the formula for $g = 0$, the two middle terms on the right side are new. It is beyond the scope of this book to prove this important theorem. Full proofs can be found in [Griffiths], [Kendig 2], [Kunz], [Springer], or [Walker], for example.

Here are some consequences of the Riemann-Roch theorem.

• When $g = 0$, the Riemann-Roch theorem's formula reduces to

$$L(-\mathfrak{a}) = \deg[\mathfrak{a}] + 1.$$

To show this, note that the two new terms in the Riemann-Roch theorem's formula are $L(\mathfrak{a}-\mathfrak{c})$ and g. Since $g = 0$, we need show only that $L(\mathfrak{a}-\mathfrak{c}) = 0$. This is true since if any divisor $\mathfrak{b}$ has positive degree, then there can be no function cutting out this divisor. If there were, the function would have additional poles to keep the number of poles equal to the number of zeros. Those additional poles imply points outside $\mathfrak{b}$ having negative coefficients, but all points of S outside $\mathfrak{b}$ have coefficient zero. We therefore assume $\deg[\mathfrak{a}] \geq 0$ so that $L(-\mathfrak{a}) \neq 0$. Since $g = 0$, $\deg[\mathfrak{c}] = 2g - 2 = -2$, so $\mathfrak{a} - \mathfrak{c}$ has positive degree and therefore $L(\mathfrak{a} - \mathfrak{c}) = 0$.

• We can deduce from the Riemann-Roch theorem that if the Riemann surface S is a torus, then there exists no function on S having a single simple pole. To show this, let P be any point on S, let $\mathfrak{a} = P$, and let's see what the formula gives for $L(-\mathfrak{a})$. The first term of the formula's right-hand side is $\deg[\mathfrak{a}]$, which is 1. The next term is $L(\mathfrak{a} - \mathfrak{c})$. Now

$$\deg[\mathfrak{a} - \mathfrak{c}] = \deg[\mathfrak{a}] + \deg[-\mathfrak{c}];$$

$\deg[\mathfrak{a}] = 1$ and $\deg[-\mathfrak{c}] = 0$ since $\deg[\mathfrak{c}] = 2g - 2 = 2 - 2 = 0$, so $\deg[\mathfrak{a} - \mathfrak{c}] = 1 > 0$, and $L(\mathfrak{a} - \mathfrak{c}) = 0$. Finally, since $g = 1$, the last two terms of the formula cancel, so that for $\mathfrak{a} = P$, we have $L(-\mathfrak{a}) = 1$. But the constant functions on S already fill out a vector space of complex dimension 1. Any nonconstant function would expand it to dimension 2, so there can be no function on S with P as a pole. Since every meromorphic function on any compact Riemann surface assumes each value the same number of times, there is no meromorphic function on a torus that assumes *any* value exactly once.

• The Riemann-Roch theorem allows us to determine the dimension of the vector space of those functions cutting out any canonical divisor. To find this dimension, choose $\mathfrak{a}$ to be $\mathfrak{c}$. The formula then reads

$$L(-\mathfrak{c}) = \deg[\mathfrak{c}] + L(\mathfrak{c} - \mathfrak{c}) - g + 1.$$

Now $\deg[\mathfrak{c}] = 2g - 2$, and $\mathfrak{c} - \mathfrak{c}$ is just the divisor with all zero coefficients, so it corresponds to the 1-space of constants on S. Substituting gives $L(-\mathfrak{c}) = (2g - 2) + 1 - g + 1 = g$. Therefore the vector space of meromorphic functions cutting out the canonical divisor has dimension g.

• The term $L(\mathfrak{a} - \mathfrak{c})$ is nonzero only when the divisor $\mathfrak{a}$ has degree equal to or less than $2g - 2$. This term becomes 0 when $\mathfrak{a}$'s degree is more than $2g - 2$, for then the divisor $\mathfrak{a} - \mathfrak{c}$ has positive degree, and by the observation made a moment ago, there are no functions cutting out $\mathfrak{a} - \mathfrak{c}$. Therefore whenever $\deg[\mathfrak{a}] > 2g - 2$, the Riemann-Roch theorem becomes

$$L(-\mathfrak{a}) = \deg[\mathfrak{a}] - g + 1.$$

• The embeddings mentioned in Chapter 5 (p. 127) have an analog for compact Riemann Surfaces that are abstractly defined, with no mention of surrounding space. The Riemann-Roch theorem guarantees that on any such surface, there are always enough functions to embed it in $\mathbb{P}^3(\mathbb{C})$. See [Fischer], p. 169.

Bibliography

[Abbott] Timothy Good Abbott, *Generalizations of Kempe's Universality Theorem*, 2008. (Online: search for "Generalizations of Kempe's Universality Theorem")

[A-S] M. Abramowitz and I. Stegun, *Handbook of Mathematical Functions*, Dover Publications, New York, 1972.

[Bliss] Gilbert Bliss, *Algebraic Functions*, Dover Publications, New York, 1966.

[B-K] Egbert Brieskorn and Horst Knörrer, *Plane Algebraic Curves*, Birkhäuser Publishing, Boston, 1986.

[Cartan] Henri Cartan, *Elementary Theory of Analytic Functions of One or Several Complex Variables*, Addison-Wesley, Reading, Mass., 1963.

[Clemens] C. Herbert Clemens, *A Scrapbook of Complex Curve Theory*, Graduate Studies in Mathematics, Volume 55, American Mathematical Society, Providence, R.I., 2001.

[Coolidge] Julian Lowell Coolidge, *A Treatise on Algebraic Plane Curves*, Dover Publications, New York, 1959.

[Fischer] Gerd Fischer, *Plane Algebraic Curves*, Student Mathematical Library, Volume 18, American Mathematical Society, Providence, R.I., 2001.

[Fulton] William Fulton, *Algebraic Curves: An Introduction to Algebraic Geometry*, 2008. (Online: search for "Fulton Curves Book")

[Griffiths] Phillip A. Griffiths, *Introduction to Algebraic Curves*, Translations of Mathematical Monographs, Volume 76, American Mathematical Society, Providence, R.I., 1989.

[Hartshorne] Robin Hartshorne, *Algebraic Geometry*, Graduate Texts in Mathematics, Volume 52, Springer-Verlag, New York, N.Y. 1977.

[H-C] Adolf Hurwitz and Richard Courant, *Allgemeine Funktionentheorie und Elliptische Functionen; Funktionentheorie*, Grundlehren 3, Springer-Verlag, Heidelberg, 1922.

[Kendig 1] Keith Kendig, *Conics*, Dolciani Mathematical Expositions, Volume 29, Mathematical Association of America, Washington, D.C., 2005.

[Kendig 2] ———, *Elementary Algebraic Geometry*, Graduate Texts in Mathematics, Volume 44, Springer-Verlag, New York, N.Y. 1977; second edition, Dover Publications, New York, 2011.

[Klein] Felix Klein, *On Riemann's Theory of Algebraic Functions and their Integrals*, Dover Publications, New York, 1963.

[Kunz] Ernst Kunz, *Introduction to Plane Algebraic Curves*, Birkhäuser Publishing, Boston, 2005.

[M-S] Robert Messer and Philip Straffin, *Topology Now!*, Classroom Resource Materials, Mathematical Association of America, Washington, D.C., 2006.

[Miranda] Rick Miranda, *Algebraic Curves and Riemann Surfaces*, Graduate Studies in Mathematics Volume 5, American Mathematical Society, Providence, R.I., 1995.

[Needham] Tristan Needham, *Visual Complex Analysis*, Oxford University Press Inc., New York, 1997.

[Picard] Émile Picard, *Traité d'Analyse*, Gauthier-Villars et fils, Volumes I–III, 1891–1896. (Online, search for "Picard Traite d'Analyse")

[Reid]	Miles Reid, *Undergraduate Algebraic Geometry*, London Mathematical Society Student Texts 12, Cambridge University Press, Cambridge, U.K., 1988.
[Seidenberg]	Abraham Seidenberg, *Elements of the Theory of Algebraic Curves*, Addison-Wesley, Reading, Mass.,1968.
[Springer]	George Springer, *Introduction to Riemann Surfaces*, 2nd edition, AMS Chelsea, 2001.
[S-S]	Lynn A. Steen and J. Arthur Seeback, Jr., *Counterexamples in Topology*, Dover Publications, New York, 1978 edition.
[van der Waerden]	B. L. van der Waerden, *Modern Algebra Volume I*, Frederick Ungar Publishing Co., New York, 1953
[Walker]	Robert Walker, *Algebraic Curves*, Dover Publications, New York, 1962.
[Whitney]	Hassler Whitney, *Complex Analytic Varieties*, Addison-Wesley, Reading, Mass., 1972.

INDEX

abelian group, 165, 180
abelian variety, 165
affine curve, 40, 46
affine curve in $\mathbb{R}^2$, 40
alpha curve, 34, 52, 55, 93, 95, 100, 106, 111, 117, 118, 121–123, 126–129
analytic branch, 52, 58
analytic continuation, 80
analytic function element, 80

Bézier cubic, 22
Bézout's theorem, 63, 66, 67, 71, 72, 100, 122, 130, 132, 135, 165, 180
birational equivalence, 118, 119, 121, 122, 124–126, 144, 152–154, 156, 157, 166
birational transformation, 118, 138, 153
blowing up a point, 137
branch parametrization, 58

Cauchy-Riemann equations, 88
compact Riemann surface, 143, 144, 146, 147, 149, 155, 156, 172–174, 178–180, 182–184
complex affine curve, 46
complex potential, 173
complex projective curve, 46
complex work, 173, 179
conformally equivalent, 155, 160, 166
conformally equivalent Riemann surfaces, 146, 153
conjugate parametrization, 56, 58, 62
connected, 75–78
connected, simply, 81
cusp curve, 55, 93, 95–97, 106, 111, 113, 114, 137
cuts out a divisor, 180

degree of a monomial, 1
degree of a polynomial, 1
degree of an algebraic curve, 1
desingularize, 121, 153, 155
differential, 175, 177–179
differential, exact, 175
differential, meromorphic, 181
discriminant, 78, 90
discriminant point, 78, 79, 81–84, 89
disk model of $\mathbb{P}^2(\mathbb{R})$, 32
divisor, 179, 180
divisor, canonical, 181
divisor, principal, 180
double line, 2
doubly periodic function, 163

elliptic curve, 158, 160, 163, 167

Fermat curve, 26
fractional power series, 53
function element, 80
function field, 118–123, 125, 127, 145, 152–154, 156, 157, 163, 164, 167, 172, 176–179, 182
fundamental domain, 160, 162, 164, 166–168
fundamental parallelogram, 164
Fundamental Theorem of Algebra, 47, 100
Fundamental Theorem of Calculus, 176
generic point, 121
genus, 102, 103, 106–110, 112, 114, 115, 117, 145–147, 149, 151, 153, 156–158, 166, 169, 172, 173, 178, 181–183
genus formula, 88, 91, 110, 114, 159

hemisphere model of $\mathbb{P}^2(\mathbb{R})$, 36

189

highest degree part of a polynomial, 24, 25
homogeneous, 38
homogenization, 38
homogenization of a set, 38
homothetic lattices, 161

Implicit Function Theorem, 81, 86
inherited topology, 76
initial part of a polynomial, 23
intersection multiplicity, 47–51, 56–59, 61, 63, 67, 94

Jacobian matrix, 87

Kempe's Universality Theorem, 28

Lagrange interpolating polynomial, 22
lattice, 148, 160, 161, 164, 166, 173
lifting, 127, 129, 130, 132, 136–138
line at infinity of $\mathbb{P}^2(\mathbb{R})$, 33
linkage, 28
linking number, 56
Lissajous figure, 16, 17, 19
lowest-degree part of a polynomial, 23

Milnor multiplicity, 113, 116
Milnor-Jung formula, 113–115
modular group, 162
morphing, 18
multiplicity of root, 47, 48

negative orientation, 84
Newton polygon, 54
node, 95, 100, 101, 106–110, 112, 116, 117, 123, 127, 129, 137, 138
nonsingular, 93
nonsingular at a point, 97
nonsingular curve, 86, 88
nonsingular point, 94

order of a point, 94
order of a polynomial at a point, 48, 50
ordinary singularity, 95, 96, 100, 108, 110–112, 130, 131, 133, 134
orientable, 75, 84, 85, 88
orientation-preserving map, 85

Pappus' hexagon theorem, 73

Pascal's theorem, 20, 71, 72
path, 76
pathwise connected, 76, 77, 81–83
permutation group, 83
place, 127
points with multiplicity, 47
positive orientation, 84
projective completion, 32
projective curve in $\mathbb{P}^2(\mathbb{C})$, 46
projective curve in $\mathbb{P}^2(\mathbb{R})$, 40

quadratic transformation, 138, 139

ramp, 79
rational parametrization, 122, 125, 128
real affine plane curve, 40
real projective curve, 40
real two-manifold, 84
residue, 178
resultant, 12–18, 59–63, 66, 67, 69, 90
Riemann surface, compact, 146
Riemann surfaces, conformally equivalent, 146
Riemann-Roch theorem, 155, 182–184
rose, 96, 111, 131, 133–135, 137

self-intersection, 96, 99, 128
similar lattices, 161
simply connected, 81
singular point, 94
singularity, 93, 94
singularity, order of, 94
singularity, ordinary, 95
space curve, 104, 106, 118, 127, 128, 131, 133, 136, 137
standard quadratic transformation, 139
symmetric group, 83

tangent cone, 23
Theorem of the Primitive Element, 145, 153
topological space, connected, 76
topological space, pathwise connected, 76
topology of $\mathbb{P}^2(\mathbb{R})$, 37
topology, inherited, 76
transcendence degree 1, 143, 145, 163

transitive, 83, 84
two-manifold, 84

unimodular matrix, 161

valuation, 154
valuation ring, 155
vector field, 174
vector space model of $\mathbb{P}^2(\mathbb{R})$, 36

Weierstrass $\wp$-function, 163, 164, 167, 179, 182

About the Author

Keith Kendig received his Ph.D. from UCLA in Algebraic Geometry, and subsequently spent two years at the Institute for Advanced Study. He is the author of *Elementary Algebraic Geometry* (Springer-Verlag Graduate Texts in Mathematics), and two other books, *Conics* and *Sink or Float? Thought Problems in Math and Physics*, both published by the MAA. In 2000 he received the Lester Ford Prize for his *Monthly* article "Is a 2000-Year-Old Formula Still Keeping Some Secrets?" He is currently an associate editor of *Mathematics Magazine* and is also on the editorial board of the Spectrum Series of MAA books.

Applications of Algebraic Curves and Algebraic Geometry to other fields have recently burgeoned, and Keith felt the need for an account of Algebraic Curves that offers to a broad range of mathematicians and scientists an inviting and accessible entry to the field. Both as teacher and author, he is well known for his lively expository style, copious examples and teachable illustrations.